CODE RED! CODE RED!

How Destruction of the Environment Poses Lethal Threats to Life on Earth

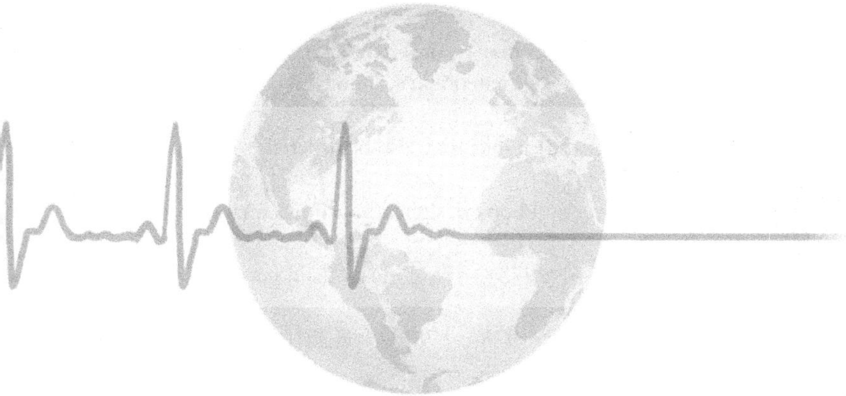

BERL FALBAUM

Cover and book designed by Anne Zimanski

Published by: S/D Books, LLC

S/D Books,
L L C

Printed in the United States

Soft cover books may be ordered by sending an email to
climatebook@yahoo.com

Soft cover books and ebooks may also be ordered from Amazon.com

Cover and book designed by Anne Zimanski

ISBN: 978-0-578-37341-6 (soft cover)

ISBN: 978-0-578-37342-3 (ebook)

About the book's title: As cited at the top of Chapter 1, United Nations Secretary-General António Manuel de Oliveira Guterres used the phrase, "Code red for humanity" after assessing a report on the environment from the Intergovernmental Panel on Climate Change (IPCC) in August 2021.

Warning

This book may be detrimental to your mental health.
It could cause depression and despair.

Contents

Introduction

For many decades, hanging on the wall of my office, I had a 22"x18" framed copy of a color photo of the Earth taken from space.

Periodically, I would gaze at the photo, marveling at the dazzling sight captured by the crew of Apollo 8 while orbiting the moon. After I retired, it hung on a wall in the family room of my house.

That photo—the first ever taken of the Earth from space—also caught the attention of former vice president, Al Gore. He discussed the photo at the beginning of his book, *an inconvenient truth, the crisis of global warming* (Rodale Books, 2006). The book was released concurrently with his Oscar-winning documentary of the same name. Gore also won the Nobel Peace Prize for his work in raising public awareness to the global warming/climate change crisis.

"We saw its [the Earth's} beauty in a new way," Gore wrote on page twelve of his book, and he opened the documentary with several breathtaking photos of Earth.

On Christmas Eve, December 24, 1968, Astronauts William A. (Bill) Anders, Frank F. Borman and James A. (Jim) Lovell, Jr., on Apollo 8, were making their fourth orbit of the moon when Anders looked out of a window and saw a spectacular sight that stunned him.

From some two hundred and forty thousand miles away, he saw a blue-white sphere: the Earth rising over the horizon. Quickly, he searched for his camera and took the color photo which was to become the most famous photo of the thousands of images taken of Earth during the decades-long space program. Ultimately, the photo was entitled, *Earthrise.*

It was hard to grasp and fully digest all of its significance. Some 7.9 billion human beings (as of early 2022), living on this awe-inspiring rotating sphere, called this home. It was and remains mind-boggling.

This is the photo, Earthrise, taken by Astronaut Bill Anders during the Apollo 8 mission in 1968. The bottom of the photo in the forefront shows the edge of the moon.

Whenever I stared at the photo in wonder, I always felt fear. I feared for the Earth's future. Can it survive? Will it survive?

Slowly, concern about the environment was being discussed publicly. The media were reporting on the dangers we faced. Politicians started addressing environmental issues. The public consciousness was awakening to the destructive behavior of humans around the world.

Yes, all to the good. Yet, I felt it was all too superficial and too late. And that was more than fifty years ago.

I did not see the kind of worldwide commitment needed to halt the potential ruination of the Earth. Whatever actions had been adopted seemed patchwork and reactive. Setting aside the financial resources which would be required, the political will did not exist.

In the late 1960s, U.S. Senator Gaylord A. Nelson, the former governor of Wisconsin (1958-62), known as the "conservation governor," was instrumental in launching what we celebrate each year as Earth Day (April 22).

On the first Earth Day in 1970, millions participated in its celebration around the country and Nelson's efforts led to the creation of the U.S. Environmental Protection Agency (EPA) that same summer. That event was followed by the adoption of much legislation designed to protect the environment.

(In 1971, I was administrative aide to Michigan's lieutenant governor, the late James H. Brickley, and helped write his speech commemorating the first anniversary of Earth Day).

In the U.S, we enacted the Clean Air Act, the Resource Recovery Act, the Occupational Safety and Health Act (OSHA), the Surface Mining Control Act, the Reclamation Act, the Toxic Substances Control Act, and many other laws. We adopted recycling and sustainability programs and other measures.

Earth Day was adopted around the world about twenty years after the United States established the day dedicated to protecting the environment. Numerous organizations were founded whose mission was to "save" the Earth and many summits/meetings have been held around the globe to deal with the complex issues involved.

The bottom line: Earth Day now receives very little attention. The Earth is in more jeopardy than it has ever been. Almost daily there are stories in the media that "we are running out of time." My fears, more than fifty years ago, regrettably, were not unfounded.

While all the actions we have taken are commendable and needed, they are not sufficient to stem the tide leading us to an uninhabitable Earth.

In my view, we are trying to stop the flow of blood from a ruptured major artery with a band-aid.

Even Gore indulges in some unwarranted optimism. While devoting one hundred and sixty pages to the catastrophic dangers faced by mankind in his book/movie, he concludes with recommendations which will do little to keep us safe. (We'll deal more with this in the closing pages of this book).

How bad is it? David Wallace-Wells, in his book, *The Uninhabitable Earth: Life after Warming* (Tim Duggan Books, 2019), provides the answer in the following first eight words of his book: "It is worse, much worse, than you think."

Moreover, most of our attention is focused on the emission of carbon dioxide (CO_2) gases from fossil fuels—coal, petroleum, natural gas and oil—which are increasing the temperature of the Earth, causing floods, droughts, fires, and hurricanes the likes of which we have never experienced before.

But the threat to the Earth as a habitable planet and the challenges we face involve much more than CO_2 gases. They include clean water shortages, threats to wildlife, overall pollution, dying coral reefs, and even garbage disposal.

Unfortunately, many of these issues do not attract headlines in the media. They are not "sexy" subjects and escape the attention they deserve.

Thus, this book, a summary, a digest if you will, of just highlights—or more accurately, lowlights—of the threats we face. This is a book for laypeople written by a layman. It discusses major issues related to global warming but also other threats which, while not directly related to warming temperatures, nevertheless, pose a threat to life on Earth.

I am neither an expert nor a scholar on the environment. Each subject I address merits a book on its own. Indeed, there are already numerous books on each subject for those who wish more detailed accounts of the issues just "touched on" here. I only cover the proverbial "tip of the iceberg." (No pun intended).

When I launched this project, my goal was to produce a book which discusses, in easy-to-understand terms, the dangers to a habitable Earth and how all life on this planet is threatened by more than just greenhouse gas emissions.

Also, I should confess, by nature and personality, I am generally a pessimist when looking at problems, usually considering the glass half empty instead of half full. The reader might take that into consideration.

I am not confident that our political, business, academic and environmental leaders have the will, finances or the needed political prowess to guide us on a path that will assure continued life on the beautiful globe that Astronaut Bill Anders captured in his photo.

b.f.

Author's Notes

To cite all the sources after every few sentences on the subjects discussed with footnote numbers and in footnotes would be annoying and onerous for readers so I have listed all the references on pages 96-99. This book is a compilation of reports, studies, facts, data and conclusions on the subjects covered that I collected by scouring the Internet as well as from books, magazines and newspaper articles.

Similarly, repeating the equivalency of Celsius to Fahrenheit was needlessly repetitive so here are the key conversions:

1.0^0 Celsius = 1.8^0 Fahrenheit

1.5^0 Celsius = 2.7^0 Fahrenheit

2.0^0 Celsius = 3.6^0 Fahrenheit

3.0^0 Celsius = 5.4^0 Fahrenheit

4.0^0 Celsius = 7.2^0 Fahrenheit

5.0^0 Celsius = 9.0^0 Fahrenheit

Celsius is used throughout the book unless a source is quoted that cites Fahrenheit.

While a ton equals two thousand pounds, a metric ton (or tonne), frequently used in this book, totals 2,204.6 pounds.

The terms "global warming" and "climate change" have different meanings. Global warming refers to the increase in global temperatures over time, resulting mainly from large concentrations of greenhouse gases in the atmosphere while climate change includes increasing temperatures and their impact on the environment, i.e., melting glaciers, changing weather patterns, raging fires, more powerful hurricanes and other disastrous consequences.

1

Global Warming/Climate Change

A code red warning, used by medical personnel in hospitals when patients face dire life-threatening circumstances, has been sounded for all of humanity living on planet Earth.

In 2021, that warning came from United Nations (U.N.) Secretary-General António Manuel de Oliveira Guterres. He used the phrase "code red for humanity" in reaction to a report from the Intergovernmental Panel on Climate Change (IPCC), a U.N. agency, which reported on unprecedented climate changes not experienced on Earth for thousands of years.

Addressing the 26th U.N. Conference of Partners (COP26) summit meeting on climate in Glasgow, Scotland in 2021, Guterres warned: "The alarm bell is deafening and the evidence is irrefutable. We are in the fight of our lives, and the fight must be won."

Once a global warming skeptic, England's Prime Minister Boris Johnson, addressing the COP26 delegates from about two hundred countries, echoed Guterres's concern, comparing the danger to the Earth to James Bond, "who generally comes to the climax of his highly lucrative films strapped to a doomsday device desperately trying to work out which coloured wire to pull to turn it off...while a red digital clock ticks down remorselessly to a detonation that will end human life as we know it."

Then he added: "It's one minute to midnight on that doomsday clock and we need to act now."

While Astronaut Bill Anders saw a beautiful rotating globe from space in 1968, the scene is much different these days as Gavin Schmidt, a senior climate advisor to the National Aeronautics and Space Administration (NASA), described in a *New York Times* photo essay in 2021. In publishing photos taken with highly technologically advanced cameras, which Anders did not have in 1968, Schmidt made these observations:

• The most obvious signs [of climate change] are in the ice. Mountain glaciers are shrinking everywhere. Whether in the high mountains of the tropics like Quelccaya in the Peruvian Andes, the Puncak Jaya Mountain on the island of New Guinea, the Alaskan ranges, the Himalayas or the Alps, we see accelerating glacier loss. In Chile, a glacier called HPS-12 retreated eight miles between 1985 and 2017.

• Sea ice has also receded drastically. As the ice recedes, less solar radiation is reflected back into space; instead, it is absorbed by the ocean, disrupting currents and ecosystems. In the Arctic Ocean in September, when the seasonal sea ice reaches its minimum, the extent of the sea ice is down forty percent on average from the 1980s.

• Melting of the great ice sheets that sit atop Greenland and Antarctica are adding enormous volumes of water to the world's oceans. Researchers have calculated that the two ice sheets together lost eighty one billion tons per year in the 1990s compared with four hundred and seventy five billion tons of ice per year in 2010, a six-fold increase. Greenland and Antarctica have lost 6.4 trillion tons of ice since the 1990s.

• In the Mississippi Delta, rising seas, compounded by sinking land and wetland degradation caused by gas and oil extraction, have resulted in a loss of land equal to the size of Delaware in the last century.

• Land isn't merely being subsumed by rising seas, but also damaged by drought, rising temperatures and wildfire. In the vast boreal regions of the Arctic, permafrost is thawing, and as it does, it upends homes, buildings and other infrastructure while releasing additional carbon dioxide and methane, a particularly potent greenhouse gas. Around Lake Baikal, in southern Siberia, the scars of fires are clearly seen from space.

(A somewhat symbolic aside: We are already "polluting" space. On several occasions, astronauts in the International Space Station had to prepare for emergency returns to Earth because they feared that orbiting debris left by nations involved in space exploration might hit and damage the space station).

We are in perilous times as warnings from the world's experts have not prompted appropriate responses. If we do not control CO_2 emissions—eighty percent of total greenhouse gas emissions—the Earth will become even warmer with devastating results.

A key statistic on carbon emissions: Around 1880, before the burning of fossil fuels, we had roughly two hundred and seventy parts per million (ppm) of carbon in the atmosphere. As of 2020, we were at 419.135 ppm, setting a new record high. The jump of 2.6 ppm over 2019 levels was the fifth highest annual increase in the National Oceanic and Atmospheric Administration's (NOAA's) sixty-three years of record-keeping.

Since 2000, the global atmospheric carbon dioxide amount has grown by 43.5 ppm, an increase of twelve percent.

The last time the atmosphere held similar amounts of carbon dioxide was during the Pliocene Period, NOAA said, about 4.1 to 4.5 million years ago.

Of the twenty-one hottest years, twenty of them were in the last twenty-five years. In 1950, we emitted six billion tons of greenhouse gases; in 2020, the total reached thirty-four billion tons, or 4.79 tons per person in the world.

We already have experienced an increase of a little more than 1° Celsius.*

Each year between 2008 and 2016 an average of 21.5 million people were forcibly displaced by sudden onset of weather-related hazards, including abnormally heavy rainfall, prolonged droughts, desertification, environmental degradation, or sea-level rise and cyclones. These people have been labeled "climate refugees."

In the United States, by 2100, large swaths of coastal land in Florida will be permanently submerged. In New York, thirty-seven percent of lower Manhattan will be at risk from storm surges by 2050, according to the Lower Manhattan Coastal Resilience Survey. Other areas in the United States, especially ocean-front cities, are in danger as well.

Climate scientists agree that it is imperative for the world to

* Rather than continually repeat "a little more than" in front of 1° Celsius, I just use 1° Celsius throughout.

keep temperatures from rising no more than 2^0 Celsius. Even if that were achieved—many experts doubt it's realistic—the consequences will be dire. Even at 1.5^0 Celsius the toll on Earth will be devastating.

As reported in *Rolling Stone Magazine*, which devoted most of its April 2021 issue to climate change, a study revealed that at 2^0 Celsius, "heat events would become 2.6 times worse, plant and vertebrate-species loss three times worse, the decline of marine fisheries two times worse, ninety-nine percent of coral reefs will die, many vulnerable and low-lying regions would become uninhabitable and the flow of refugees would rise dramatically."

Here are some other major points from the *Rolling Stone* article:

• Every ton of coal and every barrel of oil we burn heats up the atmosphere....changing currents....bringing new droughts and storms, shifting rainfall patterns, melting ice, eroding coral reefs, spawning toxic algae blooms, and moving the ocean a little closer to a world dominated by jellyfish and slime.

• Global emissions of CO_2 have increased continually since 1800. From 1850 to 2019, two thousand, four hundred gigatons* were emitted as a result of human activity. Of that, nine hundred, fifty gigatons went into the atmosphere while the rest was absorbed by oceans and land. In 2019, about 43.1 billion metric tons of CO_2 from human activities were emitted into the atmosphere, an all-time high, breaking the previous record from 2018 when these emissions grew faster than at any time since 2010-11.

The magazine did not cover the emission of methane, a green-

* One gigaton equals one billion metric tons or 2.2 trillion pounds.

house gas which is twenty-eight times more potent than carbon dioxide. The levels of methane have increased one hundred, seventy percent since pre-industrial levels.

Who are the worst offenders of carbon dioxide emissions? China is No. 1 with ten thousand and sixty-five metric tons, while the United States is No. 2 with five thousand, four hundred and sixteen metric tons.

We are dumping carbon dioxide into the atmosphere about ten times faster than volcanoes did two hundred and fifty million years ago, which cooked the planet…[and] wiped out ninety-six percent of the species on Earth and turned the ocean into a lifeless, slimy Jacuzzi.

The Jakobshavn Glacier near Greenland, the fastest-moving glacier in the world, is flowing into the sea at a rate of one hundred, fifty feet per day.

It is probably too late to alter the effect of climate change for the glaciers on Mount Kilimanjaro, large portions of the Great Barrier Reef, and Miami (yes, Florida).

What follows is a look at only a few—and I stress a few—of the major crises, not all of them directly related to global warming/climate change.

It is very likely that every community, no matter how small, is experiencing some effect of global warming/climate change/pollution.

2

Sources of CO_2 and Other Gas Emissions

Let's take a look at carbon dioxide (CO_2) emissions by economic sector in 2019:

Electricity, twenty-five percent; agriculture, ten percent; industry, twenty-three percent; transportation, twenty-nine percent; and commercial/residential, thirteen percent.

(Here are just a couple of startling facts from the transportation category: On a three-thousand mile round-trip flight, a plane emits a ton of CO_2 *per passenger*, and there are a hundred thousand take-offs and landings in the world every day).

Most of the public focus is on fossil fuels—coal, petroleum, oil and natural gas—which emit CO_2. However, there are other gases which are contributing to global warming. These include methane, CH_4, sixteen percent; nitrous oxide, N_2O, six percent; and fluorinated gases, F-gases, two percent.

• Methane is a hydrocarbon gas produced both through natural sources and human activities, including the decomposition of wastes in landfills, agriculture, and especially rice cultivation, as well as ruminant digestion and manure management associated with domestic livestock. It has a much higher warming potential than CO_2.

• Nitrous oxide is a powerful greenhouse gas produced by soil cultivation practices, especially the use of commercial and organic fertilizers, fossil fuel combustion, nitric acid production, and biomass burning (the burning of dead and live vegetation).

• Fluorinated gases are used inside of products like refrigerators, air-conditioners, foams and aerosol cans. Emissions from these products are caused by gas leakage during the manufacturing process as well as throughout the product's life.

It's worth pointing out that the meat industry is responsible for about 14.5 percent of global greenhouse (methane) gas emissions, more than the entire transportation industry.

As *The New York Times* noted: Cattle have an outsized environmental impact largely because they belch* up methane, a potent planet-warming gas.

Moreover, there is little hope in curbing these emissions because global consumption of meat is expected to rise seventy percent by 2050 compared to 2011. The National Oceanic and Atmospheric Administration (NOAA) reported in April 2022 that atmospheric concentrations of methane increased steadily over the past 15 years, and reached a new high in 2021.

Given all the sources of gas emissions, we have our work cut out for us.

---

* This fact even made it to the TV quiz show *Jeopardy!* in January 2022. Contestants were offered the following clue: If cows were a country, they'd be the sixth largest emitter of this greenhouse gas. The answer: What is methane?

3

Rainforests/Deforestation

To put it briefly: Rainforests are essential because they suck up CO_2 from the atmosphere. While the process is complicated, the bottom line is we need rainforests to stop global warming.

When forests are cut down, much of the stored carbon is released into the atmosphere as CO_2. This is how deforestation and forest degradation contribute to global warming.

Deforestation is the second largest human source of CO_2 emissions after fossil fuels, accounting for ten to fifteen percent of emissions globally.

At the global level, nineteen percent of the carbon in the Earth's biosphere is stored in plants, and eighty-one percent in the soil. In all forests, tropical, temperate and boreal together, approximately thirty-one percent of the carbon is stored in the biomass and sixty-nine percent in the soil.

One study has calculated that the world's vegetation, from the Amazon Rain Forest to Eurasian grasslands, may hold about four hundred and fifty billion metric tons of carbon, roughly equal to the carbon that humans would pump into the atmosphere over fifty years at current rates of emission.

Deforestation is responsible for more greenhouse gas emissions than all the world's cars, trucks, planes and ships combined.

Writes Daniel Howden in *The Independent*: In twenty-four hours, deforestation releases as much CO_2 into the atmosphere as eight million people flying from London to New York. Scientists have concluded that if tropical deforestation were a country, it would be the third-biggest emitter globally, ranking just below the United States and significantly higher than the European Union.

According to Global Forest Watch, between 2015 and 2017, forest-related emissions were sixty-three percent higher than the average for the previous fourteen years, rising from three billion metric tons to 4.9 billion metric tons per year.

The following is from *Time Magazine's* piece entitled, "The Amazon Rain Forest is Nearly Gone:"

"The forest stores up to one hundred and twenty million metric tons of carbon, equivalent to almost twelve years of global emissions...If cleared, much of that will go into the atmosphere. That alone would push the global climate beyond safe limits.

"With as much as seventeen percent of the forest lost already, scientists believe that the tipping point will be reached at twenty percent to twenty-five percent of deforestation even if climate change is tamed. If, as predicted by some, global temperatures rise by 4^0 Celsius, much of the central, eastern and southern Amazon will certainly become barren scrubland."

Most of the deforestation has been caused by loggers, farmers, the cattle industry and mining.

Says the Citizens for Global Solutions:

"... dams, roads, and farms blemish the Amazon Rain Forest, threatening all those that call this vast forest their home. So far, twenty percent of the Amazon Rain Forest has already been destroyed and lost forever with an estimated 2.7 million acres of forest burned each year due to human activities. Thus, the Amazon Rain Forest which once was known as the 'lung of the world,' is now an alarming producer of CO_2 due to deforestation. The destruction of this irreplaceable forest is causing the extinction of millions of species and endangering the people who live in the areas ruined by environmental and cultural devastation."

The Amazon, the world's largest rainforest, covers one thousand, five hundred and seventy-nine million acres and expands through eight countries: Brazil, Bolivia, Peru, Ecuador, Colombia, Venezuela, Suriname, and French Guiana.

The Earth's "lungs," soak up about a quarter of the world's CO_2 emissions with a total of 2.4 billion metric tons annually.

Researchers reported that over the last two decades, trees in the Amazon have been dying at an increasing rate, rendering the massive jungle a weaker absorber or "sink" of CO_2. Indeed, most of the Amazon lies in Brazil, and that part of the Amazon has become an emitter, instead of an absorber, of emissions.

(In the first few months of 2021, four hundred and thirty thousand acres of the Amazon were destroyed in Brazil alone).

But the Amazon is not the only rainforest being attacked and destroyed.

Along with the Amazon, the World Wildlife Fund (WWF) lists the following ten other "hotspots" suffering from severe deforestation: Atlantic Forest Gran Chaco; Borneo; Cerrado; Chocó-Darién; Congo Basin; Eastern Africa; Eastern Australia; Greater Mekong; New Guinea; and Sumatra.

Not on the list is Haiti where ninety-eight percent—*ninety-eight percent*—of its rainforest has been cut down.

Temple University Scientist Blair Hedges, who spent decades studying in Haiti's rainforests, reported in a 2018 paper that Haiti has only one percent of its primary forest left as various industries cut down millions of trees. And because of the deforestation, it is estimated that almost half of Haiti's species will be wiped out by the year 2035. Haiti is already undergoing a mass extinction of its biodiversity because of deforestation.

Hedges added that forty-two of the country's fifty highest and largest mountains have lost all their primary forest.

Similarly, Madagascar, the world's oldest island, has lost ninety percent of its natural forest in just four decades.

In March 2022, *The New Republic* published an article, "Crime and Forestry," on logging, much of it illegal, in Romania. The magazine reported that, "Romania is home to one of the largest and most important old-growth forests in the world. It's been referred to as the Amazon of Europe, a comparison apt and ominous in equal measure because of the speed at which it, like the Amazon itself, is disappearing."

The poet Alfred Joyce Kilmer wrote, in 1913, in his famous poem "Trees," that only God can make a tree. Kilmer could not and did not foresee how man has undone the work of God.

4

Permafrost

For most of human history, permafrost has been Earth's largest terrestrial carbon sink, trapping plant and animal material in its frozen layers for centuries.

Thawing of the permafrost is especially dangerous because it could cause a domino effect of more warming that, for now, cannot be checked by human engineering or policy.

"We now estimate the deposits contain more than 1.5 trillion tons of frozen carbon, about twice as much carbon as contained in the atmosphere," said Dr. Charles Tarnocai, Agriculture and Agri-Food Canada, Ottawa, and lead author of a study published in *Global Biogeochemical Cycles*.

Climatetippingpoints.info, a climate science blog, makes this estimate: There are one thousand, six hundred gigatons of carbon locked in Arctic permafrost.

Rising global temperatures are causing frozen Arctic soil—permafrost—in the northern hemisphere to thaw and release CO_2 that has been stored within it for thousands of years. The amount of carbon stored in permafrost is estimated to be four times greater than the combined amount of CO_2 emitted by modern humans.

Permafrost is any type of ground—from soil to sediment to rock—that has been frozen continuously for a minimum of two years and as many as hundreds of thousands of years. It can extend down beneath the Earth's surface from a few feet to more than a mile, covering entire regions, such as the Arctic tundra, or a single, isolated spot, such as a mountaintop of alpine permafrost.

The Arctic is warming twice as fast as anywhere else, and faster than it has in the past three million years. Scientists estimate there is now ten percent less frozen ground in the northern hemisphere than there was in the early 1900s. With every 1^0 Celsius of warming, an additional 1.5 million square miles of permafrost could eventually disappear.

The Arctic alone is estimated to hold nearly twice as much carbon as exists in the atmosphere now, as well as a sizable amount of methane. Estimates on how much carbon and methane may be released by thawing permafrost vary, but according to one study, as much as ninety-two billion metric tons of carbon could be emitted by 2100.

The loss of permafrost can also contribute to sea level rise. Indeed, it's estimated that if all of the Earth's permafrost thaws, sea levels could rise by as much as four inches, enough to double the flood risk for cities like San Francisco, Seattle, and Los Angeles.

Warns the journal, *Nature*: Slow and steady thawing will release around two hundred billion metric tons of carbon over the next three hundred years under a business-as-usual warming scenario. That's equivalent to about fifteen percent of all the soil carbon currently stockpiled in the frozen north.

5

Glaciers

Nearly all of the world's glaciers are losing mass at an ever increasing pace, contributing to more than a fifth of global sea level rise this century, according to various studies. If they disappear, 2.6 billion people will suffer drinking water shortages.

Melting glaciers add to rising sea levels, which in turn increases coastal erosion and elevates storm surge as warming air and ocean temperatures create more frequent and intense coastal storms like hurricanes and typhoons.

Specifically, the Greenland and Antarctic ice sheets are the largest contributors of global sea level rise. The Greenland ice sheet is disappearing four times faster than in 2003 and already contributes twenty percent of current sea level rise.

How much and how quickly these Greenland and Antarctic ice sheets melt in the future will largely determine how much ocean levels rise in the future. If emissions continue to rise, the current rate of melting on the Greenland ice sheet is expected to double by the end of the century. If all the ice on Greenland melted, it would raise global sea levels by twenty feet.

For the first time, researchers have studied all of Earth's some two hundred thousand glaciers, and discovered that between

2000 and 2019, the world's glaciers lost an average of two hundred and sixty seven billion metric tons of ice each year. That's enough water to submerge Switzerland under twenty feet of water—every single year.

Between 2000 and 2004, glaciers lost two hundred and twenty-seven billion metric tons of ice per year. But between 2015 and 2019, they lost an average of two hundred and ninety-eight billion metric tons annually.

Studies blame the United States and Canada for half of the world's glacier loss.

One study, published in *Nature,* said that the fastest-melting glaciers were situated in Alaska and the Alps.

Here is just one example of melting glaciers. This is the Lyell Glacier, the second largest glacier in California's Sierra Nevada Mountains in Yosemite National Park. It has lost sixty-two percent of its mass and one hundred and twenty vertical feet of ice in the last century, much of that in recent years. Scientists predict it will lose the rest of its ice by roughly 2030. Indeed, it is no longer considered a glacier.

But the authors also expressed concern about the retreating glaciers in the Pamir Mountains, the Hindu Kush and the Himalayas, which provide water for more than 1.5 billion people.

"During the dry season, glacial melt water is an important source that feeds major waterways such as the Ganges, Brahmaputra and Indus rivers," said Romain Hugonnet, lead study author and researcher at ETH Zurich University and the University of Toulouse.

Scientists also have found a melting ice sheet in the southwestern region of Greenland that was releasing huge amounts of mercury into the nearby rivers and fjords.

"There are surprisingly high levels of mercury in the glacial melt waters we sampled in southwest Greenland," said Jon Hawkings, a postdoctoral researcher at Florida State University and the German Research Centre for Geosciences. "And that's leading us to look now at a whole host of other questions such as how that mercury could potentially get into the food chain."

"We didn't expect there would be anywhere near that amount of mercury in the glacial water…," said Robert Spencer, associate professor of Earth, Ocean and Atmospheric Science at Florida State University.

6

Oceans

The oceans are heating up fast. The amount of heat being added to oceans is equivalent to every person on the planet running one hundred microwave ovens all day and night. Scientists have found that, in 2018, the heat in the upper part—the top six thousand, five hundred feet—of the world's oceans was the highest ever recorded since record keeping began in the 1950s.

The International Union of Conservation of Nature (IUCN) reported that the oceans have absorbed more than ninety percent of the heat generated by human activity since the 1970s. The oceans are at their warmest in one hundred and thirty-six years of records.

The Intergovernmental Panel on Climate Change (IPCC) has predicted that by 2100 the ocean will absorb two to four times more heat than between 1970 and the present if global warming is limited to 2^0 Celsius and up to five to seven times more at higher emissions. Globally, marine heat waves will very likely increase by a factor of fifty by 2081-2100.

Along with rising temperatures, the oceans are threatened by pollution which comes primarily from the following sources: wastewater and sewage; industrial waste; fossil fuels; sewer line leaks; and fertilizer/pesticide runoffs from such sources as septic tanks and farms.

Five huge patches of debris have formed in the Indian Ocean, Atlantic Ocean and Pacific Ocean. These patches vary in size but the Pacific patch, halfway between Hawaii and California, which has made headlines and is known as the Great Pacific Garbage Patch, is made up of 1.8 trillion—that's *1.8 trillion*—pieces of trash and covers an area twice the size of Texas.

The Great Pacific Garbage Patch.

Eight million metric tons (17.6 billion pounds) of plastic are dumped into the oceans each year. By 2050, there will be more plastic in the oceans than fish (measured in pounds). Plastic takes about four hundred years or longer to degrade and, worse, the process releases dangerous chemicals that further contaminate the seas.

While oil spills attract the most attention, they make up only twelve percent of the oil in the oceans. (The rest comes from a variety of sources).

More than eighty percent of the world's sewage (eighty-five percent in the Caribbean) is discharged untreated into oceans. The discharge of human waste has been labeled "ocean defecation."

The journal, *Environmental Science & Technology*, reports that rivers collectively dump anywhere from 0.47 million to 2.75 million metric tons of plastic into the seas every year.

The ten rivers that carry ninety-three percent of that trash are the Yangtze, Yellow, Hai, Pearl, Amur, Mekong, Indus, the Ganges Delta in Asia, the Niger and the Nile in Africa. The Yangtze alone dumps an estimated 1.5 million metric tons of plastic waste into the Yellow Sea.

The U.S. Environmental Protection Agency (EPA) estimates that a single person pumps one hundred and fifty thousand gallons of sewage per week.

Around the world, ocean dead zones—areas with little or no oxygen that marine mammals need to survive—are growing at an alarming rate. Most of this is directly tied to agriculture: soil or land degradation resulting from huge amounts of chemical fertilizer and livestock manure runoff which make their way down rivers and into the oceans.

A United Nations report states: "…fertilizers entering coastal ecosystems have produced more than four hundred ocean 'dead zones.'"

Scientists predict that, because of poor farming practices, the Gulf of Mexico could experience one of the largest dead zones in history: an astounding eight thousand square miles, an area the size of Massachusetts. Because of fertilizer runoffs, the Gulf will barely be capable of sustaining any form of life.

Most of the ocean garbage—seventy percent—sinks to the bottom and cannot be cleaned up.

The oceans have absorbed twenty to thirty percent of the CO_2 since the 1980s, causing acidification. Increased acidity will have significant negative effects on many organisms important in the marine food chain, including plankton and crustaceans.

Rising sea surface temperatures are also intensifying stratification between surface and sub-medium seawater, leading to the spread of ocean regions around the world in which oxygen concentrations are very low and life is difficult to sustain.

And there is noise pollution coming from ships and the military which can cause damage to invertebrates that are the food source for tuna, sharks, and sea turtles.

Countries all over the world, especially undeveloped countries that do not have comprehensive sewage systems, are dumping billions of gallons of untreated human waste into the oceans (and rivers).

Then there are the cruise ships. In just one year, they dumped one billion gallons of untreated sewage into the oceans.

The Natural Resources Defense Council (NRDC) sums it up as follows: "Our rivers, reservoirs, lakes, and seas are drowning in chemicals, waste, plastic, and other pollutants."

7

Rising Sea Levels

Based on the most recent projections, global sea levels are very likely to rise at least twelve inches above 2000 levels by 2100 even with lower emissions. At the highest level of greenhouse gas emissions, sea level rise could be as high as 8.2 feet above 2000 levels by 2100.

Global sea levels have risen by roughly eight inches over the past century and the rate of increase is accelerating. Various studies project that further increases of 1.5^0 Celsius could produce about a three-foot sea level rise by 2100, displacing about fifty-six million people in some eighty countries.

With a 1.5^0 Celsius increase, the Arctic Ocean will become ice-free in the summer about once every one hundred years and at a 2^0 Celsius increase, once every ten years. The three major causes: Ocean expansion, melting of ice sheets, especially in Greenland and Antarctica, as well as melting of glaciers.

The Solomon Islands, a country of volcanic islands and coral atolls in the southwestern Pacific Ocean, has lost five islands due to rising seas. Another six islands have lost more than twenty percent of their surface area, forcing communities to relocate.

As the flattest country on Earth, the Republic of Maldives, located in the Indian Ocean southwest of Sri Lanka and India, faces the possibility that much of its land will be underwater by the end of this century. Sea level rise is likely to worsen existing environmental stresses in the Maldives, such as periodic flooding from storm surge, and a scarcity of fresh water for drinking and other purposes.

Even with an "average" sea level rise, the Maldives is projected to experience a sea level rise of about 1.5 feet and to lose some seventy-seven percent of its land area by around 2100. With a rise of three feet, the Maldives could be almost completely inundated by about 2085.

More than three hundred thousand coastal homes, with a collective market value of about $117.5 billion, are at risk by 2045. In addition, about fourteen thousand coastal commercial properties, assessed roughly at $18.5 billion, are also at risk during that time frame.

By the end of the century, homes and commercial properties worth more than one trillion dollars are in danger. This includes as many as 2.4 million homes, the rough equivalent of all the homes in Los Angeles and Houston combined.

The properties facing risks for rising sea levels by 2045 house five hundred and fifty thousand people and contribute nearly $1.5 billion toward the property tax base. Those numbers jump to about 4.7 million people and twelve billion dollars by 2100.

The Marshall Islands, situated in the central Pacific Ocean halfway between Hawaii and Australia, are projected to lose forty percent of its buildings in the capital, Majuro, with ninety-six percent of the city facing possible severe flooding caused by climate change.

Jakarta, the capital of Indonesia, is described in studies as one of the fastest-sinking cities in the world. Government officials have announced plans to move its headquarters to Borneo, a more central and higher location.

In the U.S., states with the most homes at risk by the end of the century are Florida, with about one million homes (more than ten percent of the state's current residential properties); New Jersey, with two hundred and fifty thousand homes; and New York with one hundred and forty-three thousand homes. Resources for the Future, a nonprofit organization that conducts studies on the environment, stated that Miami "is the most vulnerable major coastal city in the world."

Around Florida, sea levels are eight inches over what they were in 1950, and they are rising at one inch every three years.

State officials are studying projects which would cost more than four billion dollars to protect sewage systems, raise roads, improve stormwater systems and construct sea walls.

In New York, the coastline is nine inches higher than it was in 1950. Like Florida, New York is considering spending four billion dollars to protect against future sea level increases.

Coastal communities around the world, including Rio de Janeiro, Osaka, and Shanghai, face the possibility of major disasters resulting from rising sea levels.

If all the glaciers and ice caps on the planet were to melt, global sea levels would rise by about two hundred and thirty feet, flooding nearly every coastal city in the world, according to the U.S. Geological Survey. The oceans would cover all the coastal cities and land area would shrink significantly.

The Intergovernmental Panel on Climate Change (IPCC) projected a high-end estimate of two feet through 2099, but its 2014 report raised the high-end estimate to about three feet. A number of later studies have concluded that a global sea level rise of six to eight feet in this century is "physically plausible." A conservative estimate of the long-term projections is that each Celsius degree of temperature increase triggers a sea level rise of approximately 7.5 feet over two millennia.

8

Coral Reefs

Coral Reefs are in crisis, dying at an alarming rate worldwide. An estimated twenty-five percent of coral reefs have already disappeared and an estimated two-thirds of all coral reefs are at risk.

Coral reefs will continue to decline by a further seventy to ninety percent at 1.5^0 Celsius and ninety-nine percent at a 2^0 Celsius increase.

Called the "rainforests of the sea," coral reefs are the greatest expression of ocean life, and the most biodiverse ecosystem on Earth with thirty of thirty-four known animal phyla present.

By midcentury, every reef in the world will be eroding away. In some fifty years, these reefs, which have been around about two hundred and fifty million years, will be "crumbling ruins."

More than fifty percent of the world's coral reefs have died in the last thirty years and up to ninety percent may die within the next century. Very few pristine coral reefs still exist.

More than 2,838,240 acres of coral reef are lost every year, totaling an area ten times the size of New York City.

Nearly ninety percent of new coral in Australia's Great Barrier Reef, the world's largest coral reef system and considered one of the seven wonders of the natural world, has collapsed in the past few years. *National Geographic Magazine* declared the reef half dead.

Ocean life diversity has dropped sixty percent in Indonesian reefs (where the majority of the world's coral exists) due to climate change and pollution.

If carbon dioxide gas emissions are not reduced, the oceans may be so acidic that some corals that had been able to withstand these conditions may erode quicker than they can rebuild.

The reefs in Florida, home to the last living reef in the United States, are in "dire" condition, according to scientists, due to rising temperatures and chemical dumping. Over the last forty years, coral reefs around Florida and throughout the Caribbean have become degraded. Since 1975, more than ninety percent of the reefs in the Florida Keys have lost their living coral cover, according to one study by Dr. Phil Dustan, a professor at the College of Charleston.

Coral reefs provide a three-dimensional habitat for a variety of plants and animals. While they occupy less than one percent of the ocean floor, coral reefs support more than twenty-five percent of all marine life. They also shelter shorelines from storms and erosion, and provide food and jobs for coastal communities dependent on tourism and fishing.

An estimated eighty-eight percent of the reefs in Southeast Asia, the most species-rich reefs on Earth, are at risk.

Scientists say climate change, disease, bleaching, overfishing, pollution and ocean acidification are decimating these fragile ecosystems and putting communities and livelihoods in jeopardy.

"Coral reefs have been in decline worldwide—I think that's pretty commonly accepted," says Tyler Eddy, a research scientist at Memorial University of Newfoundland and author of a major study. "We didn't necessarily know the magnitude of how much, when we looked on a global scale, that reefs had declined."

In addition to finding that half of living corals have died since the 1950s, researchers discovered that coral-reef-associated biodiversity dropped by sixty- three percent. Healthy reefs support thousands of different corals, fish and marine mammals, but bleached reefs lose their ability to support many species.

And the study showed that the loss of coral species wasn't equal across reefs. Some corals are proving more sensitive than others, leading biologists to worry that vulnerable coral species will be lost before they can be documented or preserved.

Coral reefs, along with sea grass beds, mangrove habitats and other tropical marine environments, support the highest marine biodiversity in the world.

More than five hundred million people worldwide depend on them for food, storm protection, jobs, and recreation. Their resources and services are worth an estimated three hundred and seventy-five billion dollars each year, yet they cover less than one percent of the Earth's surface.

9

Clean Drinking Water

Water is the blood and plasma line for life on this planet. And that line is being severely threatened.

Sure, seventy-one percent of the Earth is covered by water. But the problem is ninety-seven percent of that is saltwater. Of the remaining three percent, 2.5 percent is fresh water but that's frozen in ice, primarily in Antarctica and Greenland, leaving a mere .5 percent for drinking water.

(Eight countries, Brazil, Russia, the United States, Canada, Colombia, China, the European Union and Indonesia have most of the world's fresh water reserves).

Global demand for water is expected to increase by twenty to thirty percent by 2050 and by that year, the United Nations says that more than half of the world's population will live in water stressed regions where demand will exceed supply.

Warmer temperatures, along with decreasing rainfall, lead to faster and greater evaporation, affecting lakes, reservoirs and rivers. Scientists report that one third of the world's largest aquifers ranging from China, India and the United States, are being depleted and are seriously threatened.

Two billion people, that's one in four people, don't have a safely managed water service at home. About eight hundred million people—one in ten people—do not have access to clean water. The United Nations Children's Fund (UNICEF) has reported that more than five thousand children under five die daily from diarrhea caused by water-related diseases.

Almost four hundred and ninety million people in the world—one in sixteen—have no choice but to drink dirty water.

Here are the top ten countries with percentages of their population that lack access to clean water (defined as having access to water under a thirty-minute trip):

Mozambique, fifty-three percent; Niger, 54.2 percent; Chad, 57.5 percent; Democratic Republic of the Congo, fifty-eight percent; Angola, fifty-nine percent; Somalia, sixty percent; Ethiopia, sixty-one percent; Uganda, sixty-one percent; Papua New Guinea, sixty-three percent; and Eritrea, 80.7 percent.

About four billion people experience extreme water shortages for at least one month a year.

Almost two billion people—one out of every four—drink water from sources that have some levels of fecal contamination.

As to rivers, here are the sixteen most polluted rivers in the world: Neva, Yamuna, Yangtze, River Tame, Jordan, Tijuana, Matanza-Riachuelo, Buriganga, Marilao, Sarno, Citarum, Doce, Yellow, Ganges, and two in the U.S, the Mississippi and the Cuyahoga in northeast Ohio. Three of those rivers, the Yangtze, Yellow and Ganges in the Delta, are among the ten that carry ninety-three percent of the trash into the oceans as mentioned on page 21.

We might note that the "mighty" Mississippi carries an estimated 1.5 million metric tons of nitrogen pollution into the Gulf of Mexico annually, creating a "dead zone" in the Gulf about the size of New Jersey.

The United States faces water problems of its own as experts warn that decreasing precipitation and increasing populations could produce unprecedented water shortages.

For instance:

Experts have sounded an alarm that within fifty years, many regions of the United States could see their freshwater supply reduced by as much as a third. With water basins in trouble, water shortages could impact most of the United States, including the central and southern Great Plains, and the southwest and center Rocky Mountain states. California, the South and the Midwest could also suffer similar water shortages.

Federal officials warned there is a one in three chance that Lake Powell, on the Arizona-Utah border, will fall so low that the Glen Canyon Dam will be unable to produce electricity.

The water crisis in the southwestern part of the United States may spread to other areas in the country when local waterways are unable to replenish their resources to meet ongoing and increasing demands.

A water shortage was declared, for the first time ever, at Lake Mead, one of the Colorado River's main reservoirs. Lake Mead, as of 2021, contained twelve million acre-feet (one acre-foot equals three hundred and sixty thousand gallons) of water, far below its capacity of almost thirty million acre-feet. The last time it was full was about twenty years ago. The lake supplies water to twenty-two million people, and it is feared it may soon be dry.

The Colorado River, which has flowed one thousand and four hundred miles from the Rockies to the Gulf of California for millions of years, no longer reaches the sea.

Obviously, as the U.S. population grows, demand for water will increase. On average, each American uses eighty to one hundred gallons of water daily, with a total usage of about three hundred and forty-five billion gallons. By 2100, the U.S. population is projected to increase by two hundred million to about five hundred and fourteen million, putting much more stress on water supplies.

Then there are the Great Lakes, the largest freshwater system in the world, with more than twenty percent of the world's fresh water, and the primary water source for more than forty million people.

Nearly twenty-two million pounds of plastic are dumped into the Great Lakes annually, with more than half going into Lake Michigan, the second largest freshwater body in the world but the most polluted of the Great Lakes. Lake Ontario gets 3.2 million pounds of plastic; Lake Erie, 5.6 million pounds; Lake Huron, 1.4 million pounds; and Lake Superior, 70,500 pounds.

The statistics cited above are just a few on the scarcity of clean water and the pollution of water supplies. The scenario, in a word, is chilling.

How in the world will the world be able to clean up the mess that already exists, let alone supply clean water as demand grows and warming temperatures threaten existing supplies?

10

Fish

Fish are being harmed by the warming of the oceans' temperatures and dying coral reefs and they are also endangered by overfishing.

Since the 1970s, the climate crisis has made oceans warmer and more acidic, reducing the number of fish we rely on for food and putting the future of fish in peril, according to a United Nations report.

An estimated seventy percent of fish populations are fully used, overused, or in crisis as a result of overfishing and warmer waters. Ultimately, present fishing practices could lead to the extinction of some species and unless the world adopts remedial policies, one study went so far as to suggest there will be no fish left by 2050.

Rising temperatures mean oceans have less oxygen and more acidification, forcing fish to seek new habitats, creating large dead zones where life cannot survive. There are more than four hundred coastal dead zones already with many other areas in danger of being free of fish.

Warmer waters also lead to algae blooms, red tides (harmful algal bloom which produce toxic and harmful effects on humans, fish and birds) that further decrease oxygen.

With a 1.5^0 Celsius increase, one global model projects a decrease in the global annual catch for marine fisheries by about 1.5 million tons; at a 2^0 Celsius increase, three million tons.

As to overfishing, several environmental organizations have expressed concern about the status of worldwide fishing. They fear that millions of tons of fish are being hauled out of oceans faster than they can be naturally replenished.

The United Nations Food and Agriculture Organization (FAO), in an overview entitled, "Snapshot of the Global Situation," stated: Of the six hundred marine fish stocks monitored by the FAO, three percent were underexploited, twenty percent were moderately exploited, fifty-two percent were fully exploited, seventeen percent were overexploited, seven percent were depleted, and only one percent was recovering from depletion.

Other facts from experts give further reasons for concern:

• Almost thirty percent of fish stocks commercially fished are overfished.

• In the Northeast Atlantic and nearby seas, thirty-nine percent of fish stocks are classified as overfished. In the Mediterranean Sea and the Black Sea, studies found that of eighty-five stocks, eighty-eight percent are overfished.

• More than one billion people worldwide depend on fish as their primary source of protein.

• Approximately ninety percent of stocks of large predatory fish are already gone, as overfishing has targeted the largest fish at the top of the food chain.

• Commercial fishing from the world's lakes and rivers has quadrupled in the last fifty years and now totals 8.7 million tons annually.

• Production of farmed fish has increased from twenty-four million tons in the mid-1990s to sixty-seven million tons in 2012.

Scientists warn that, "we are at a tipping point...there is no guarantee there will be recovery by mid-century." It's not impossible, they maintain, "but we need to be quite cautious."

11

Climate Refugees

The environmental crisis has added a new term to our vocabulary: Climate refugees. While the scientific community differs on the definition, generally these are people who have had to abandon their homes because of disasters—hurricanes, rising seas, famine, droughts, etc. Many of these are caused, of course, by climate change.

The following are data from various agencies on the plight of climate refugees.

• The International Refugee Assistance Project (IRAP) reported that since 2008 environmental disasters have displaced three times more people than violence and armed conflict.

• The Institute for Economics and Peace (IEP), an Australian think tank, projects that 1.2 billion people could be displaced by climate change events by 2050. The major countries and areas at greatest risk are: Sub-Saharan Africa, South Asia, the Middle East, North Africa, Afghanistan, Syria, Iraq, Chad, Pakistan, Ethiopia, Iran, Haiti, India and China.

• The World Bank has predicted that one hundred and forty-three million people in just three areas—Sub-Saharan Africa,

South Asia and Latin America—will be forced to move because of disasters.

• In 2019, the five countries with the most climate refugees were: India, five million; Philippines, four million; Bangladesh, four million; China, four million; and the United States, nine hundred and sixteen thousand.

• In 2017, 68.5 million people worldwide were displaced, more than at any point in human history, says the Brookings Institution.

• While future impacts of climate change will disproportionately affect the world's poorest, mass migrations of refugees will also create economic, financial and political pressures on countries around the globe.

• A study by Columbia University climate researchers published in the journal, *Science,* projected that if global temperatures continue their upward march, applications for asylum to the European Union could increase two percent to nearly four hundred and fifty thousand per year by 2100.

• Over the past thirty years, the number of people living in coastal areas at high risk of rising sea levels has increased from one hundred and sixty million to two hundred and sixty million, ninety percent of whom are from poor developing countries and small island states.

• Since 2008, an average of twenty-four million people has been displaced by catastrophic weather disasters each year. As climate change worsens storms and droughts, climate scientists and migration experts expect that number to rise.

• Meanwhile, climate impacts that unravel over time, like desert expansion and sea level rise, are also forcing people from their homes.

• *Inside Climate News* published the following headline: "By 2050, 200 Million Climate Refugees May Have Fled Their Homes."

• A majority of the 30.7 million people displaced in 2020 were fleeing floods, wildfire, drought or heat waves, according to one report by the International Federation of Red Cross and Red Crescent Societies. And these trends are only likely to increase in the next decades.

My research on rising seas and climate refugees led me to a story on the Yup'ik community in Newtok, Alaska, which is being destroyed by rising seas. The community of about three hundred and fifty residents has witnessed an accelerated thawing of permafrost, causing building foundations to deteriorate and more than one hundred feet of Newtok's coast has broken off into encroaching waters. Predictions are that Newtok will be completely lost within a few years. The Yup'ik people have lived in Newtok for a millennium.

With the belated help of the U.S. government, the entire community was moved to a new village, Mertarvik, about nine miles upriver from Newtok, at a cost of more than one hundred million dollars.

"Climate change" may be a new term, but the outlook is that it will have a permanent place in our vocabulary.

12

Wildlife

If you have read this far and are not depressed or unmoved by any of the facts, perhaps the following will create a little despair: One million animal and plant species are now threatened with extinction, many within decades, more than ever before in human history.

This dire prediction comes from the prestigious U.N.-backed panel: Intergovernmental Science-Policy Platform on Biodiversity and Ecosystem Services (IPBES).

The report stated unequivocally: "Nature is declining globally at rates unprecedented in human history."

The IPBES report, published in 2019, is considered one of the most comprehensive and authoritative ever. It is the first intergovernmental report of its kind. Compiled by one hundred and forty-five expert authors from fifty countries over the past three years, with input from another three hundred and ten contributing authors, the report assesses changes over the past five decades, providing a comprehensive picture of the relationship between economic development and its impact on nature. It also offers a range of possible scenarios for the coming decades.

It cites global warming as one of the major culprits. The others are invasive species; changes in land and sea use; urban sprawl; direct exploitation of organisms; and pollution. Also on the list is poaching and illegal wildlife trading, a multi-billion dollar business.

More than three hundred and twenty species have died out over the past millennium. Estimates are that about thirty-two percent of the remaining animal populations are dwindling. With reductions of about ten percent, insect species are threatened as well.

The average abundance of native species in most major land-based habitats has fallen by at least twenty percent, with most of that decline since 1900. More than forty percent of amphibious species, almost thirty-three percent of reef-forming corals, and more than a third of all marine mammals are threatened.

The numbers of invasive alien species per country have risen by about seventy percent since 1970, across twenty-one countries.

Some scientists envision this threat to wildlife as leading to the sixth mass extinction of species in the history of the world, *but the first one caused by humans.*

Just one specific example: In 1971, Kenya* had a population of eleven million and in 2021 Kenya reported a population of 53.7 million. In 1971, it had one hundred and sixty thousand elephants, and twenty thousand black rhinos. In 2021, those numbers dropped to thirty-five thousand elephants and one thousand black rhinos and only two—that's two—northern

* I used Kenya as an example because I had the opportunity to go on a photo safari to Kenya in 1998. It was an unforgettable experience.

white rhinos, and they are both female. Lions, cheetahs, hyenas and wild dogs are all endangered.

As one study put it, "some [animals] are killed by new roads and power lines." And, I am sure, the decrease in the animal population will continue with forecasts predicting that Kenya's population will reach ninety-five million by 2050. This same scenario probably plays out in many countries.

(As of this writing in 2022, Kenya was suffering from a devastating drought caused by climate change with millions at risk of starving. Carcasses of wild animals, which died of hunger and thirst, pockmarked the entire country).

Most of the attention is generally given to large mammals such as polar bears, Asian elephants, tigers, rhinos, chimpanzees, and leopards. But that paints a false and incomplete picture because almost all land animals are affected by global warming, as are whales, sharks, sea turtles, penguins, and other marine life.

The following is from *The New Yorker*, November 2021: "Our world is in the midst of a crisis of biodiversity. The U.N. estimates that '…we keep speeding extinction …by converting more and more of the world's natural spaces into human ones. Already, we significantly changed three-quarters of the planet's land and two-thirds of its oceans, squeezing out untold numbers of wild creatures.'"

In addition to increased rates of disease and degraded habitats, climate change is also causing changes in species themselves which threaten their survival. Rising temperatures have led to ecological changes.

Climate change has affected habitats, making them, at times, inhospitable to animals. This causes animals to change breeding habits and forces them to migrate, which endangers their chances of survival.

A quick look on how birds and bees have been affected:

The Cornell Lab of Ornithology reports that the North American bird population is down by 2.9 billion breeding adults, with devastating losses among birds in every biome (flora and fauna). Forests alone have lost one billion birds. Grassland bird populations collectively have declined by fifty-three percent, or another seven hundred and twenty million birds.

BirdLife International provides the following facts: one hundred and fifty-nine species are extinct; two hundred and twenty-three are critically endangered; four hundred and sixty endangered; seven hundred and ninety-eight vulnerable; and one thousand and one near threatened.

Bees?

A study by the journal, *One Earth*, found that in recent decades, the number of bee species reported in the wild has declined globally. The sharpest decrease—about twenty-five percent—occurred between 2006 and 2015.

There are eighty to one hundred million managed beehives globally and all are experiencing dramatic declines.

Here is the bottom line on bees: They are, of course, responsible for and vital to the world's food supply. States Petpedia: Without bees we would have no food.

How pervasive is the threat to animals from human behavior?

I offer the following just as a "minor" example.

Manatees in Florida are dying in record numbers. In 2021, more than one thousand died because of a lack of food. The cause, a loss of seagrass—down by about ninety percent—resulting from waste and fertilizer run-offs from farms and lawns.

To help save the manatees, state officials are providing food—cabbage and lettuce—even though they recognize that feeding wildlife can produce new problems. But, they acknowledge, it is a desperate last-ditch effort.

Meanwhile, the Everglades in Florida are in danger. Humans have altered the natural flow of water throughout the Everglades. The result is an ecosystem in collapse. Since the 1800s, water diversions and flood-control projects have severed the flow of water in the Everglades, while large areas have been converted to agricultural or residential areas.

The update, which includes assessment of 105,732 animal and plant species to date, lists more than twenty-eight thousand species as threatened with extinction, and attributes much of the declines to human exploitation.

Out-of-sight means out-of-mind and that is especially true in assessing the threats to the smallest life forms in nature. There is little, if any discussion, on the impact on insects and microbes, which are essential to the "circle of life."

Consider: With a 1.5^0 Celsius increase in temperature, six percent of insects, eight percent of plants and four percent of vertebrates are projected by 2100 to lose more than half of their climatically determined geographic range. At a 2^0 Celsius increase, these percentages double or triple.

Edward O. Wilson,* the world renowned Harvard University sociobiologist, recommended in two of his books, *The Future of Life* (Knopf, 2002) and *Half-Earth: Our Planet's Fight for Life* (Liveright, 2017) that we dedicate fully half of the Earth's surface to nature and its wildlife.

Wilson believed there was still time to save the Earth. Basically, he argued that humans could set aside half of the Earth primarily for the millions of species that inhabit the planet with humans and save about eighty percent to ninety percent of the species, which would bring the extinction level down to what it was before the arrival of humanity.

I think it's fair to ask without being disrespectful: How likely is it that this will happen?

* Dubbed "Darwin's natural heir," winner of two Pulitzer Prizes and author of more than thirty books, Wilson died at ninety-two in December 2021.

13

Plastics

As was pointed out in the section on oceans, by the year 2050, there will be more plastic than fish (measured in pounds) in the oceans. One metric ton of plastic enters the ocean every four seconds. Plastic pollution has increased tenfold since 1980.

The New Yorker published a story in 2020 on Victor Lance Vescovo, an American private equity investor, retired naval officer, and undersea explorer who has developed a submarine to explore the deepest parts of the oceans, depths never reached by man before. On one such trip nearly to the ocean floor, about thirty- six thousand feet, Vescovo spotted a plastic bag and candy wrappers. His crew reported that almost every sample taken from the sea's bottom had been contaminated with microplastics (plastic particles five millimeters or smaller).

Eight million metric tons of plastic are dumped into the ocean each year, washing up on previously pristine parts of the planet and are a threat to the survival of more than six hundred species of marine wildlife.

Buried in landfills, the typical plastic trash bag takes one thousand years to degrade, giving off toxins as it does. By 2050, we'll have twenty-six billion metric tons of plastic waste.

Studies show that synthetic threads, which flow into wastewater treatment plants from washing clothes, account for a significant portion of the microplastic pollution problem. Experts estimate that between 1950 and 2016, the amount of synthetic microfibers entering the world's oceans, rivers, lakes, and soils totaled about 5.6 million metric tons. This issue has received very little, if any, attention.

But synthetic threads from washing clothes aren't the only problem. In addition to synthetic fabrics, plastic can enter the soil from car tires that degrade on roadways, from cosmetics that wash into wastewater and concentrate in sludge, and from tools that farmers use on their land, such as plastic mulch, seed coatings, and coverings that suppress weeds and pests.

The Norwegian Institute for Water Research, which coauthored an Op-Ed article for *Environmental Science & Technology*, stated that as much as three hundred thousand tons of microplastics end up on North American farmlands each year.

Plastic manufacturing in the first ten years of this century eclipsed the total produced in the entire last century. Now we're producing—and discarding—almost three hundred and sixty million metric tons a year.

What's more, chemicals added to plastics are absorbed by human bodies. Some of these compounds have been found to alter hormones or may lead to other serious health problems. People are exposed frequently to chemicals from plastics through the air, water, food and use of consumer products.

Plastic debris, laced with chemicals and often ingested by marine animals, can injure or poison wildlife. Floating plastic waste, which can survive for thousands of years in water, also serves as mini-transportation devices for invasive species, disrupting habitats.

Around four percent of world oil production is used as a feedstock to make plastics, and a similar amount is consumed as energy in the process.

More than one hundred and eighty species of animals have been documented to ingest plastic debris, including birds, fish, turtles and marine mammals, according to one report. Plastic trash has been found in more than ninety percent of the world's sea birds, and in the stomachs of more than half of the world's sea turtles and whales.

Plastic debris is found absolutely everywhere, from the Arctic to Antarctica. It clogs street drains in our cities; it litters campgrounds and national parks; and is even piling up on Mount Everest.

While some recommend more recycling of plastic, not all plastic is recyclable and not all recyclable plastic is recycled. Only 8.7 percent of all plastic waste was recycled in 2018 in the United States.

People are often confused by the terms, "break down" versus "biodegradable" (or "compostable"). When plastics are broken down, this simply means one large piece of plastic is reduced into a bunch of smaller pieces of plastic. These smaller pieces of plastic can be consumed by smaller animals, but are still indigestible.

A minority of plastic is "compostable" or "biodegradable," which means they can be reduced to their chemical components. Other plastics can only be successfully composted by industrial or municipal facilities after first being separated from other, non-biodegradable, plastics.

Around the world, one million plastic drinking bottles are purchased every minute, while five trillion single-use plastic bags are used worldwide every year. In total, half of all plastic produced is designed to be used only once and then thrown away.

According to the U.S. Environmental Protection Agency (EPA), in 2018, Americans generated 35.7 million tons of plastic in the United States. Some 12.2 percent of that makes up the total municipal solid waste generated.

And the future does not look promising. Since the 1950s, the rate of plastic production has grown faster than that of any other material. We've also seen a shift away from the production of durable plastic, and towards single-use plastics.

More than ninety-nine percent of plastics are produced from chemicals derived from oil, natural gas and coal, all of which are dirty, non-renewable resources. If current trends continue, by 2050 the plastic industry could account for twenty percent of the world's total oil consumption.

Estimates are that by the year 2030, some fifty-three million tons of plastic will exist in our oceans, rivers, and lakes, despite global efforts to reduce plastic waste.

With most plastic piled up in landfills or incinerated (giving off harmful gases), one obvious step would be to discourage buying disposable plastics, and to encourage recycling when it can be done effectively.

14

Garbage Disposal

Eventually, we will run out of space to dispose of the trillions of pounds of garbage the world produces year after year.

While out of the limelight and not as deadly as CO_2, greenhouse gas emissions or deforestation, garbage disposal nevertheless presents a serious environmental issue.

Humans generate more than two billion metric tons per year, according to the World Bank. Global garbage is expected to reach 3.4 billion metric tons annually by 2050. About twelve percent of all municipal waste in 2016 was plastic—two hundred and forty-two million metric tons of it.

The United States, China, Brazil, Japan and Germany are the leading trash generators. The United States generated two hundred and sixty-two million metric tons of solid waste in 2015.

Labs produce lots of waste, some of it toxic. One study estimated that every year the plastic waste from laboratories could cover an area twenty-three times the size of Manhattan ankle-deep.

Some of this trash leaks dangerous greenhouse gases that contribute to climate change. Solid waste landfills are the third-largest source of methane gas emissions in the United States, according to data from the U.S. Environmental Protection Agency (EPA).

In 2019, landfills released fifteen percent of methane gas emissions, which is equivalent to emissions from more than 21.6 million passenger cars driven for one year.

Recycling helps but much of the garbage cannot be recycled. The United States, for instance, recycles only about thirty-five percent of its garbage, while poorer countries average about four percent. Some, of course, don't recycle at all.

The United Nations Environment Programme (UNEP) predicts the amount of waste will probably double in lower-income African and Asian cities as a result of population growth, urbanization and rising consumption.

Pollution produced by garbage runs into rivers and seeps into groundwater. Flooding is caused by garbage clogging drains, and the atmosphere can be poisoned by the toxic discharge from trash.

When waste is not collected, the frequency of illnesses such as diarrhea doubles and acute respiratory infection—linked to the burning of waste—is six times higher, according to health experts.

Many rich countries like the United States, Japan and Germany have resorted to exporting their garbage, particularly to Asia. But Asian countries are refusing to accept this trash. These countries faced major problems of how to deal with tainted garbage that couldn't be easily recycled.

"Typically, seventy percent of a shipment can be processed,

and the other thirty percent is contaminated with food," said Thomas Wong, manager of Impetus Conceptus Pte., Ltd., a Singaporean company that shreds locally produced plastic waste before sending it to recycling mills in Malaysia and Vietnam.

When China banned imports in 2018, other countries followed suit. Shipments were diverted to Southeast Asia, which soon became overwhelmed, forcing countries to stop the influx. Malaysia banned imports and then Thailand stopped issuing import licenses while the Philippines sent garbage back to Canada. Indonesia tightened waste-import rules after discovering shipments containing toxic waste. India and Vietnam also adopted restrictions. Garbage disposal may not make it into the headlines but, eventually, the world will have to face the problem and find new ways to dispose of its trash.

Whatever the solutions, we can be confident they will not solve the problem completely and they will be costly.

15

Population Growth

The preceding pages were the good news. Now comes the bad.

If a world of 7.9 billion people faces such a potentially dismal future—and the subjects covered here hardly begin to tell the full story—then what can we expect when the world adds another billion or more people?

The question, of course, answers itself. Even if you are an optimist, you have to be, at least, a little concerned.

The world's population at the time this was written (early 2022) was 7.9 billion. When Gore released his movie/book in 2006, it stood at 6.6 billion. This was an increase of 1.3 billion or 19.6 percent in just fifteen years.

All studies are in general agreement that the Earth's population will reach 9.1 billion by 2050—that's less than thirty years away, and an increase to almost eleven billion by the end of the century. Some predict an even larger increase.

The United Nations has estimated the planet will need twice as much food by 2050 than it produced when this was written. The United Nations Food and Agriculture Organization

(FAO) estimated that we will need to increase world food production by sixty to seventy percent to feed nine billion people. Along with food, of course, there will be increasing demands for land, water and energy.

Edward O. Wilson, the prominent Harvard University sociobiologist I mentioned earlier, warned in his studies that the Earth has only the maximum capacity to support nine to ten billion people.

Between 2013 and 2015 the world population increased about sixty percent in the developing world vs. two percent in higher income countries; the corresponding food needs will not be uniform either.

The following is from a U.N. report: Since 1970, the global human population has more than doubled (from 3.7 to 7.9 billion), rising unevenly across countries and regions.

Global energy consumption, for instance, is expected to grow by forty-eight percent in the next twenty years and consumers will meet that increase by, primarily, non-renewable energy sources, i.e., gas, oil, petroleum and coal.

Nature, in an article entitled, "Options for Keeping the Food System within Environmental Limits," reported, "without targeted changes pressures on various environmental systems will increase fifty to ninety percent by 2050 compared to 2010."

The magazine said that the Earth has limits—"planetary boundaries"—that can't be exceeded without potential dire consequences. These boundaries, which involve climate change, loss of biodiversity, deforestation, atmospheric aerosols (smog), stratospheric ozone depletion and a supply of fresh water, define a safe operating space for humanity.

The Center for Biological Diversity adds:

"As the world's population grows, so do its demands for water, land, trees and fossil fuel—all of which come at a steep price for already endangered plants and animals."

And here is the most unjust news of all: Poor countries, already feeling greater pressures than richer ones for food, clean water and energy, will suffer even more as the demand for basic needs increases in direct relation to population growth.

The NATO Review reported under the headline, "Population Growth, the Defining Challenge of the 21st Century":

"Without taking action now, billions of people across the world will face thirst, hunger, slum conditions and conflict in response to droughts, food shortages, urban squalor, migration and ever depleting natural resources, while capacity tries to catch up with demand.

"And the projected growth in demand is staggering. More mouths to feed and changes in tastes for food types will mean:

"---Doubling agricultural production in four decades;

"---Water consumption will need a thirty percent increase by 2030; and

"---Three billion extra people will seek shelter in urban settings by mid-century."

The report continued: "Add to this the need for energy to underpin economic growth in post-industrial, industrial and newly industrializing nations alike, with demand doubling by 2050, and there is a significant challenge for governments and society at large."

Daniel Altieri observed in an article entitled, "The Effects of Overpopulation on Water Resources and Water Security" for Swarthmore College Environmental Studies:

"As the global population is expected to keep growing in the coming decades, the negative impact that humans will have on Earth's finite resources, especially water, will become increasingly apparent as areas of the world will start to experience drastic shortages of water, leading to instability in food production, industry, social order, and political and military control.

"In order to limit the amount of chaos and conflict that will ensue over limited water resources, there needs to be compromise and cooperation between all countries, not just the nations that are water stressed, to provide water management techniques, newer and more efficient technology to conserve as much water as possible, and strict security and enforcement of all regulations to prevent groups and individuals using water to gain power."

Finally, Population Media Center writes:

"Slowing down, stopping and eventually reversing human population growth—these are ethical imperatives that will help improve the chances for future generations establishing living scenarios with the planet.

"The most ethical gift we can give the people and creatures of the last 21st century and early 22nd century is a chance. In the realm of population, this means working across the generations to ensure a much smaller, more ecologically right-size number of humans."

Thus, population control *must* be part of the solution and I discuss that in the next chapter.

16

What Can We Do?

Let's answer that question by accepting the fact that we can't repair much of the damage we have done to ourselves. It is irreversible.

We can't "refreeze" glaciers which take one hundred to one hundred and fifty years or more to fully form, nor can we refreeze permafrost that has melted.

We can't reforest rainforests. The rainforests that have been destroyed remain destroyed.

We can't restore the pristine quality of the planet's waters. Some contamination will endure in oceans, lakes, rivers even if we were to stop all pollution immediately which, of course, is impossible.

We can't clean up all the soil in the world which has been contaminated by fertilizers and other chemicals.

We can't rebuild communities that have been submerged in rising seas.

We can't bring back animals that we drove to extinction.

We can't fix many other problems we have created through the years and we need to accept the reality of irreversibility in many instances and deal with them.

As I indicate in the Introduction, this short book only summarizes the crises confronting the Earth. I only devoted a few pages to each while experts and scholars have examined these subjects at great length in scientific papers and books.

Not touched on at all are such "challenges" as: poverty, air pollution,* genetic modification, ozone layer depletion, mining, natural resource depletion, natural disasters, nuclear waste,† acid rain, farming,‡ light and noise pollution, and urban sprawl.

It's clear that everything we have done or tried in the last fifty years or so has not worked. The Earth is getting warmer, pollution is becoming more insidious each year, life-threatening weather is becoming more frequent, and fires are destroying more forests than ever before, even in areas that had never experienced a forest fire.

---

* Air pollution is one of the world's leading risks for death, killing seven million people annually. About ninety percent of the world is exposed to unhealthy air.

† With the disposal problem of nuclear radioactive waste unresolved, more than a quarter of a million metric tons of highly radioactive nuclear waste sit in storage in nuclear power plants and weapons production facilities worldwide with more than ninety thousand metric tons in the U.S alone. The U.S. Department of Energy reports that U.S nuclear reactors produce more than two thousand metric tons of radioactive waste a year and, lacking disposal locations, remain on sites of companies that produced the waste. This waste remains radioactive for thousands of years.

‡ For instance, *60 Minutes* aired a segment in December 2021 on how climate change has devastated the wine industry in much of Europe. But, other countries, such as England and even Norway, which historically were unable to grow grapes because of cold temperatures, are now producing wine because their temperatures have risen.

Addressing the issue of fires specifically, the United Nations published a one-of-a-kind report in February 2022 that predicted wildfires would increase by up to fifty-seven percent by the end of the century because of climate change. The report was the work of fifty researchers from six continents

Neither the Paris Climate Agreement (COP21)) nor the 26th Conference of Partners (COP26) give us hope that more will be done in the future.

The PCA, adopted in December 2015, called for countries:

• To attempt to limit global temperature increases to 1.5^0 Celsius, and to keep them "well below" 2^0 Celsius above pre-industrial times.

• To limit greenhouse gas emissions from human activity to the same levels that trees, soil and oceans can absorb naturally between 2050 and 2100.

• To set their own emission-reduction targets which are to be reviewed every five years.

The PCA also asked rich countries to help poorer nations by providing funds—"climate finance"—to help them adapt to climate change and switch to renewable energies.

Six years later in 2021, COP26 delegates agreed to a "phase-down of unabated coal power," which is the major emitter of CO_2 emissions; called for new rules for trading carbon credits across borders and adoption of new, more ambitious targets to curb emissions; and asked member nations to report annually on their efforts summarizing their work to reduce emissions.

There were positives: the United States and China pledged to work in concert on climate issues; more than one hundred nations adopted the goal to reduce methane gas emissions by thirty percent while one hundred and thirty countries promised to work toward halting deforestation. Finally, major financial institutions are to adjust their portfolios toward the goal of attaining net-zero emissions by 2050.

As stated earlier in the book, COP26 did not call for an end of the use of coal. Alok Sharma, COP26 president who presided over the summit, held back tears when, at the last minute, the delegates voted to "phase down" rather than "phase out" the use of coal.

Also, the U.S. did not sign the pledge, apparently because it did not want to anger U.S. Senator Joseph (Joe) Machin III (D-West Virginia) who had close ties to the coal industry. (China, India and Australia did not sign the pledge either). With the Senate divided 50-50 between Democrats and Republicans, the Biden administration, needing every single vote to pass its legislative proposals, did not want to risk losing Manchin's vote.

The COP26 pact also failed to assure proper financing to attain the goals. The United Nations Environment Programme (UNEP) has estimated that developing countries alone would need seventy billion dollars a year to comply and enact climate-saving programs, and that figure will double by 2030. COP26 also expects developed countries to compensate poorer countries for irreparable damage caused by climate change.

And here is the major weakness: The agreements depend on *voluntary* action by member countries. There are no penalties for maintaining the status quo, for doing nothing. Thus, all the provisions agreed to might best be described as a "wish list."

But even as a "wish list", it is sad that the PCA and COP26 accept an increase to 1.5^{0} Celsius rather than reducing the temperature from the 1^{0} Celsius already experienced. If we are suffering from rising seas, raging wildfires, historic droughts, horrendous and more frequent hurricanes, and other devastating consequences from the present rise of 1^{0} Celsius from pre-industrial levels, we are sure to face an even more cataclysmic future at 1.5^{0} Celsius.

By focusing on a $1.5\ ^{0}$ Celsius increase, we imply that, if achieved, we would solve the world's environmental crisis, and that it ignores all the other catastrophic challenges the world faces.

The media routinely use 1.5^{0} Celsius as the goal, seemingly endorsing it, without any reflection on what that means to global warming/climate change and the future of life on Earth.

We have resigned ourselves to deal with the many crises without committing to try and lower the temperatures. We use the word "reduce" rather than "eliminate" global warming which means the threat to our very existence will continue.

For instance, if I pour water into a glass at the rate of an inch an hour and then "reduce" it to an inch every two hours, the glass will still fill up eventually and ultimately overflow.

As a layman, I was not sure I was correct in this conclusion. So, I asked some experts if my interpretation was accurate. Unfortunately, I was told I was right.

What follows is a summary of what I was told:

> *We should be aiming to limit warming to 1^{0} Celsius, or even reduce it to 0^{0} Celsius. While 1.5^{0} Celsius is signifi-*

cantly worse than 1⁰ Celsius, it is significantly better than 1.6⁰, 1.7⁰, etc. No one in the scientific community has stated that it's technically possible to limit warming below 1.5⁰ Celsius in the foreseeable future because each additional tenth of a degree in warming is already "baked in" so there is little we can do about it. Thus, 1.5⁰ Celsius has come to represent a best-case scenario—albeit one which will be catastrophic—and that we seem increasingly unlikely to achieve. Some predict an increase of 4-5⁰ Celsius.

As an example of the frustrations a concerned layperson has when seeking information on such issues, here is one experience:

I contacted the COP21 (the Paris agreement) and COP26 press offices by email and asked to speak to an official about the apparent acceptance of a 1.5⁰ Celsius increase, but I never heard from either organization.

I followed up with another email to both containing the following question: Why is the goal to keep temperatures below 1.5⁰ Celsius when we are already suffering disastrous consequences at an increase of 1⁰ Celsius from pre-industrial levels? Shouldn't the goal be to keep it at 1⁰ Celsius or, better yet, below 1⁰ Celsius?

I received an answer from Stephan Ernest, of the press office for the United Nations Framework Convention on Climate Change (UNFCCC) under whose auspices both summits were held. He wrote:

"It is the parties who decide the aims and provisions of the Paris agreement. The UNFCCC secretariat is devoted to supporting the intergovernmental negotiating process on climate change. Thus, we are not in a position to provide an adequate response on behalf of the Parties to the Convention." He also suggested that I consult the "official document" from COP26.

Sadly, but not unexpectedly, Ernest dodged the question. I understand that it will be up to the parties of the conventions to decide how to proceed. Indeed, as I write earlier in this chapter, that is one of the defects of the agreement. Compliance is voluntary.

I had read both the COP26 and COP21 pacts, and I wanted to know why they both discuss 1.5^{0} Celsius as a goal instead of lower temperatures. But Ernest avoided the point of my question.

As to possible solutions, even Al Gore, the former vice president who has done outstanding work on raising the public's conscience about what we face, has fallen into the "we can do this" trap.

In his book/movie, *an inconvenient truth, the crisis of global warming,* he devotes most of his attention to the crises haunting humanity. He concludes:

"We are witnessing an unprecedented and massive collision between our civilization and the Earth. We are trashing the planet."

Then, briefly at the end of the movie/book, he cites "solutions" such as wind power, solar energy, fluorescent light bulbs, green roofs, hybrid cars and other similar programs.

Yes, taken together these measures can reduce—and I emphasize "reduce"—our reliance on fossil fuels. But they won't affect the increase in temperature which is crucial to maintaining a livable

Earth. These measures will not keep us from reaching 1.5° Celsius or, more likely, higher temperatures. No one disputes that.

After meticulously explaining how we are destroying Earth, Gore tells his audiences that they can help stem the climate warming tide. He is giving them false hope, diluting how serious the crisis is.

"Each one of us is a cause of global warming and each of us can become part of the solution…," he writes in his book and in his narration in the movie.

For instance, he urges people to ride a bike or walk to work instead of riding in a car, to remember to turn off the lights they aren't using, and not let the door of their refrigerators remain open too long.

Come on, Mr. Vice President.

Admittedly, these are "feel good" measures, to be sure. But even if the entire world adopted all his recommendations, it would do little—no, let's just say it—it would do nothing to stem global warming and the resulting climate change or the threats facing the Earth.

Gore also suggests that consumers "avoid over-packaged food and other products with extra wrappers and layers of unnecessary plastic."

The next time you are in a supermarket take a minute to scan the thousands of items. Anything not packaged in a tin can has some plastic component whether it is the packaging or a label. We will never rid ourselves of plastic.

Of course, there is nothing wrong with making people feel good and advising them they can play a role in solving the global warming/climate change crisis. But there is a difference between giving people hope and giving them false hope that negates the seriousness of the issue.

True, wind power, solar panels, recycling and electric cars have laudable benefits, but they also bring their own threats to the environment, threats which have generally been ignored.

Wind power and solar panels combined require at least ten times as much land per unit of power produced than coal or natural gas-fired power plants.

Expanding wind and solar by ten percent annually until 2030 would require, as some have pointed out, a chunk of land equal to the state of South Dakota, according to Princeton University and an analysis by *Bloomberg News*.

This land requirement would inevitably cause invasion of wildlife habitats. Wind turbines, in particular, are dangerous to the bird population, while solar panels are constructed with hazardous materials.

Much has been made about the bright promise of electric cars. But they (like cell phones and tablets) require nickel for batteries. This demand has led to more mining for nickel all over the world, particularly, in pristine areas.

In an attempt to spur an increase in the supplies of nickel, Elon Musk, co-founder and CEO of Tesla, Inc., the world's largest manufacturer of electric vehicles, publicly promised lucrative contracts to mining companies. He added, "Please mine more nickel."

In December 2021, NBC aired a special on how one of the world's largest nickel mines is expanding on Palawan Island in the Philippines, threatening its rainforests and "putting one of the most beautiful places on Earth in peril." The network called the island an "ecological wonder."

NBC Correspondent Jacob Ward ended his story with the oxymoronic observation: "...clean energy comes with an environmental cost." That suggests that Ward believes we need to destroy the environment by adopting programs designed to save it.

The New York Times made the same point in December 2021 when it published a page one story on nickel mining in Goro, New Caledonia. In the first few paragraphs, the reporter, Hannah Beech, described the pristine landscape, writing it was "one of the most biodiverse on the planet,...astonishingly beautiful" until she came upon Goro's nickel mine which provided a "different vista," one of "a gouged red earth pierced by belching smokestacks and giant trucks rumbling across the lunar-like terrain."

The headline? "Can a Dirty Nickel Mine Fuel Tesla's Green Ambitions?"

We should also recognize the work of many non-profit organizations which engage in commendatory efforts in protecting the Earth. But raising endangered animals on a farm, while praiseworthy, does little to mitigate the killing of thousands of endangered species in the wild.

Like local, state and national governments, non-profit organizations don't have the finances to launch the kinds of efforts that are needed.

Banning plastic straws sounds enticing but does virtually nothing to curtail the disposal of trillions of tons of plastic into our waterways and landfills.

Urging the public to plant trees will not exhume from the ashes the irreplaceable rainforests so vital to life on Earth.

Finally, there is recycling, arguably, the most popular "save the world" program.

As with other proposed solutions, recycling brings huge problems of its own. These include major production costs; unsafe and unhealthy sites; some materials produce pollutants; and sites have been found to emit methane, a gas which is more damaging than CO_2.

This is not to say that we should abandon these efforts. But we should work to improve them and eliminate inherent risks.

There are two steps which could help even though they too may be politically and financially unrealistic: Population control complemented by a worldwide effort called "direct air capture" (DAC).

First, population control.

As discussed on pages 57-61, projections are that by 2050, the world population will stand at 9.1 billion. So, even if we were successful with all the projects on the table now, the Earth will be stretched to its limits to support another 1.2 billion people.

If we could reduce (and here the word "reduce" works because it does not just reduce increases but eliminates pressures) the number of humans on Earth, we might ease environmental challenges.

Let's say we set a goal of reducing the population by one billion (to 6.9 billion) or even more from 2022 levels of 7.9 billion. At a reduction of one billion, in 2050 when we are projected to have 9.1 billion, we would have about 2.2 billion fewer people and that could make a difference.

That would mean pressure on Earth would be dramatically reduced. We would need less food, clean water, energy, and all the resources needed for humans to survive. It would give the Earth some breathing room.

It's true that in the late 1970s and early 1980s China tried controlling its population by limiting families to one child each. It didn't work and China, in 2016, raised the limit to two children and then to three in 2021.

But given China's population of 1.4 billion, even at one child per family, the overall world population would have increased.

Now to DAC. This program seems to make sense. It involves sucking greenhouse gases out of the atmosphere, not reducing the increases but eliminating the gases from the air.

Nineteen countries have plants that remove CO_2 from the air. In the process, machines suck in ambient air and push it through a filter. The captured CO_2 is injected into the soil and, in about two years, according to experts, turns into stone.

The largest DAC plant, known as Orca, in Iceland, captures about four thousand metric tons annually, equal to taking eight hundred and seventy cars off the road. The problem: In 2019 alone, we emitted more than forty billion metric tons— that's billions—of CO_2 into the air. Those levels were higher than at any time in at least four million years. Quora, a Q&A website in Mountain View, California, estimates we have three thousand, one hundred and forty billion metric tons of CO_2 in the atmosphere.

Then there is the cost of DAC. Estimates range from as low as one hundred dollars a metric ton to one thousand, one hundred dollars a metric ton. But even at the lower end of the scale, the financial requirements are astronomical and, probably, prohibitive for the kind of effort needed to make a difference.

However, if we were to combine population control with a comprehensive worldwide affordable, and I stress "affordable" program, to suck CO_2 out of the air—not just reducing emissions but eliminating CO_2—that could make a difference, a huge difference. Again, elimination, not reduction should be the goal.

In 2021, interestingly, President Joe Biden signed an infrastructure bill which included $3.5 billion for four DAC plants. The four plants, when completed, would be able to suck one million metric tons of CO_2 from the atmosphere.

The IPCC put it very bluntly: Global warming below 1.5^0 Celsius is impossible without large-scale removal—note the word "removal"—of CO_2.

There are some who maintain that even under present circumstances the Earth will recover because it is resilient. But resilience depends on stopping the attacks. And that is nowhere in sight.

17

Conclusion

My objective in this relatively brief book was to offer summaries of just some of the major environmental issues facing mankind to provide laypeople with a picture of how life on Earth is hanging by a thread, a thin and weak thread.

Where does all of this leave us? In trouble, deep trouble.

As *The New York Times* reported on its front page at the end of COP26 in 2021:

"Even if countries fulfill all the emissions promises they have made, they still put the world on a dangerous path toward a planet that will be warmer by some 2.4^0 Celsius by the year 2100, compared to preindustrial times."

Thomas L. Friedman, *The New York Times* columnist who has written, for many years, about environmental dangers facing the world, followed up with:

"I have nothing against Glasgow [COP26]. But we will not decarbonize the global economy with a lowest-common-denominator plan...Not possible."

There is little to give us hope that in the future mankind will be more successful in preserving this planet. Indeed, China, the No. 1 polluter of greenhouse gas emissions into the atmosphere, at about thirty percent (10,065 metric tons), did not even attend the summit. (Neither did Russia).

The United States is No. 2 in CO_2 emissions at fifteen percent (five thousand, four hundred and sixteen metric tons) followed by India, seven percent (two thousand, six hundred and fifty-four metric tons), Russia, five percent (one thousand, seven hundred and eleven metric tons), and Japan, three percent (one thousand, one hundred and sixty-two metric tons).

COP26 President Sharma observed at the end of the conference that while pledges to keep the goal of limiting warming to 1.5^0 Celsius are "within reach ...[but] its pulse is weak."

Nathaniel Rich, in his book, *Losing Earth: A Recent History* (MCD Books, 2019), documents how beginning in 1979, the world had a chance of making a difference in controlling global warming/climate change during the following decade. But it failed.

The book discusses the birth of climate denialism and the fossil fuel industry's coordinated effort to thwart climate policy through misinformation, propaganda and political influence.

We have no evidence that anything has changed since Rich's groundbreaking work.

There are other reasons to be pessimistic about any progress. Consider:

- The 2015 Paris Climate Agreement (COP21), signed six years before the 26th Conference of Partners (COP26) pact, agreed to a similar commitment and nothing was achieved.

• While COP26 was the 26th climate summit meeting, the situation has only worsened every year.

• 2020 was the second warmest year on record, and it came just .02^0 Celsius short of tying 2016 for the warmest year ever recorded.

• 2019 was among the top three warmest years ever.

• The ten warmest years have all occurred since 2005 and seven of the ten have occurred since 2014.

The National Oceanic and Atmospheric Administration (NOAA) concluded that even "if annual emissions increase more slowly and begin to decline significantly by 2050, models project temperatures would still be at least 2.4^0 Fahrenheit [1.32^0 Celsius] warmer than the first half of the 20th century and possibly up to 5.9^0 Fahrenheit [3.2^0 Celsius] warmer."

In addition, the Climate Transparency Report, touted as the world's most comprehensive annual review of progress made by members of the G20* on climate action, found in its October 2021 report that while the use of renewable sources of energy increased, the performance of member countries was insufficient to keep temperatures from reaching 1.5°Celsius or higher.

Thus, it should be evident by now, to all parties, that what

* The members of the G20 are: Argentina, Australia, Brazil, Canada, China, France, Germany, India, Indonesia, Italy, Japan, Republic of Korea, Mexico, Russia, Saudi Arabia, South Africa, Turkey, the United Kingdom, the United States, and the European Union.

has been tried and what we are trying has not worked and is not working. All the "investigative" reports paint a very stark picture, a very dismal one.

Without change, the "new normal" means the acceptance of vicious storms, uncontrollable wildfires, unending droughts, violent hurricanes, and continual rising seas.

In accepting the inevitability of global warming, people acquiesce to living with a succession of catastrophes which, in turn, lead to increased poverty, devastating famines and economic turmoil.

In the face of such stoic resignation, should we continue to recycle, support sustainability programs, replace "gas guzzlers" with electric cars, and limit our use of plastics when possible? Should we bother to walk instead of drive, and conserve water when washing/bathing? Of course. But as we do these things, we must also perfect these measures to get the most out of them and avoid serious repercussions.

Admittedly, taken together such steps give us psychological comfort that comes from doing something. But, unfortunately, these measures are meaningless given the monumental problems that we face. And we should not kid ourselves about that. The evidence is irrefutable; we have failed and failed miserably.

Greta Thunberg, the teenage Swedish environmental activist, critical of how the world is not taking the necessary actions, protested with her followers outside COP26 in Glasgow, shouting: "No more blah, blah, blah."

Some ask: How much time does the Earth have before it's too

late? No one has answered that question definitively but the deadline is somewhere before the turn of the century. Some believe it is too late already, depending on one's definition of "too late."

One measurement, a so-called Climate Clock that sits about ten stories above New York City's Union Square in lower Manhattan, projects that 1.5⁰ Celsius may well be reached in less than ten years.

Developed by artists Gan Golan and Andrew Boyd and created by a fifteen-member team of environmental activists and supported by numerous organizations, the clock, eighty feet across, computes continually how much time we have left—down to seconds—before we exceed 1.5⁰ Celsius.

In the photo below, taken April 19, 2021, the Climate Clock shows that, under present conditions, we have six years, two hundred and fifty-six days, twenty-two hours, zero minutes and fourty-three seconds remaining to prevent global warming rising above 1.5⁰ Celsius.

Courtesy of 350.org for Climate Clock

One of the factors used in the Climate Clock is called the "carbon budget." According to the Intergovernmental Panel on Climate Change (IPCC), the remaining carbon budget before reaching 1.5⁰ Celsius is five hundred billion metric tons of carbon dioxide gas emissions from 2020 onward.

During 2020-2021, we emitted close to eighty billion metric tons which means we have four hundred and twenty billion metric tons left in the budget after 2021.

When we deplete the budget, it is expected that to be the year global temperatures reach 1.5⁰ Celsius.

And here is more bad news: *Forbes Magazine* reported on studies that "promises" made by world leaders "are wildly off the mark."

The magazine said that current policies could put the world on track for warming at 2.7⁰ Celsius by 2100. Even if pledges to control warming were initiated, we would nevertheless heat the planet by 2.4⁰ Celsius.

Given the prospectus that emerges from these pages, mankind's ultimate question is: What kind of future can we expect for the Earth and humanity?

The scientific community warns that unless we find comprehensive solutions, we will have millions of climate refugees, more frequent major droughts and famines, food and clean water shortages, and coastal areas around the world being submerged by rising seas.

There are denialists. A small minority disagrees with assessments like those made in this book and elsewhere.* These denialists claim the scientists are all doomsayers, "alarmists."

Yet, in every poll taken during the last ten years or so, no matter how the questions were posed, more than ninety percent of climate and environmental scientists agreed that the world is in peril.

Can anything be done? Perhaps. If—and it's a gigantic "if"—we were able to create universal unanimity on what needs to be done. Let us hypothesize that the scientific community were charged to develop solutions to the crises and countries around the world agreed to implement the dictates of this committee's findings and recommendations.

The Natural Resources Defense Council (NRDC), in August 2021, made the point on the need for a global consensus, stating, two months before the COP26 summit, that to keep the temperature increase to 1.5^0 Celsius, "Every government at every level—national, state, city, town—every business sector, every private enterprise, every individual must be in alignment. Whatever steps can be taken to reduce greenhouse gas emissions from burning coal, oil and gas, and preserve...its oceans, its forests, its soils, its permafrost, must be taken immediately and decisively."

"Negative trends in nature will continue to 2050 and beyond in

* In 1998 and 2007, some thirty-one thousand scientists signed a petition, stating there was "no convincing scientific evidence that human release of carbon dioxide will, in the foreseeable future, cause catastrophic heating of the Earth's atmosphere." The scientific community dismissed the petition, saying no documentation was provided and questioned the credentials of the signers, arguing many were not experts, indeed, did not even work, in the fields of ecology, climate or the environment.

all of the policy scenarios explored..." stated the IPBES report quoted earlier in this book.

But is it realistic to expect countries to put aside politics and their special interests and for industries to abandon their profit motives to create a kumbaya moment?*

We can find the answer easily. All we need do is review how well humanity, through the ages, has addressed such problems as famines, poverty, insidious diseases, racism, discrimination, the plague of war.

Even if the adjectives "unlivable" and "uninhabitable" were defined with well-meaning, albeit, unrealistic optimism, we face catastrophic consequences.

It is time to change course; indeed, it is long overdue. The carrot approach has not worked; we need to use the stick. Given humanity's failures in the past, appropriate jurisdictions might adopt *mandatory* policies with such severe punishments that potential violators will be reluctant to not obey more restrictive requirements.

Even with new mandatory regulations, we need to recognize the politics of the issues will be overwhelming. We can expect unrelenting push-back from such industries as coal, transportation, logging, mining and cattle. What's more, the costs will be enormous, in the hundreds of *trillions* of dollars.

Most important, underdeveloped countries will feel the effects

---

* It is worth noting that while greenhouse gas emissions in the United States decreased in 2020 largely because of a variety of restrictions during a coronavirus pandemic, emissions increased by 6.2 percent the following year when coal plants roared back to normal operations. So much for progress.

much more drastically than those who have the economic power to fight what appears to be a catastrophic onslaught.

Steve Killelea, founder and executive chairman of the Institute for Economics and Peace (IEP), said:

"Over the next thirty years, lack of access to food and water will only increase without urgent global cooperation. In the absence of action, civil unrest, riots and conflict will most likely increase."

In December 2021, *The New York Times* published a "video editorial" online—"Postcards from a World on Fire"—which reported at least one major effect of climate change in all of the one hundred and ninety-three member countries in the United Nations. The depth and breadth of the project was outstanding, but, as with this book, it only touched the surface.

In introducing the piece under the headline, "We've Failed Our Planet. This is an SOS," Kathleen Kingsbury, opinion editor, wrote, "…taken together, [the articles] tell a story about what we consider to be the most existential issue facing the planet today. As the video editorial that is at the core of the project says: 'Open your eyes. We have failed. The climate crisis is now.'

"…[C]limate change is already underway. At this point, we can't stop it. But we can work to limit the damage."

I agree with Kingsbury that we are helpless in altering the course of climate change, and I doubt that we can even manage it.

The environment is facing disasters on several fronts. The evidence is in. The data are irrefutable. The science is irrefutable. We are marching toward an "unlivable," "uninhabitable" world.

In late February 2022, the IPCC published its latest report on climate change, stating, "Human-induced climate change is causing dangerous and widespread disruption in nature and affecting the lives of billions of people around the world, despite efforts to reduce the risks."

"This report is a dire warning about the consequences of inaction," said Hoesung Lee, Chair of the IPCC.* "It shows that climate change is a grave and mounting threat to our wellbeing and a healthy planet. Our actions today will shape how people adapt and nature responds to increasing climate risks."

Two points need to be made about the report:

• It centers primarily on global warming and not several other issues which threaten the planet, some of which are discussed in these pages.

• It predicts dire consequences if global warming increases by 1.5° Celsius, a standard which appears to have been deemed acceptable by the world.

Code red warnings have been sounded over several decades, at numerous international climate summit meetings and by several sectors in the world community. If only we had paid attention earlier and taken the appropriate actions, actions which we are still not prepared to adopt and, even if implemented, given the damage done, would do little to reverse the deadly course we are on.

---

* On April 4, 2022, as this book was going to press, the IPCC issued another report, reiterating its warnings about global warming. "Without immediate and deep emissions reductions across all sectors, limiting global warming to 1.5°C is beyond reach," said Lee. "We are at a crossroads. The decisions we make now can secure a liveable future. We have the tools and know-how required to limit warming,"

Some six decades ago, the biologist/author Rachel Carson, in the dedication of her classic book *Silent Spring* (Houghton Mifflin, 1962) which documented how pesticides were destroying the environment, quoted the philosopher/theologian/writer Albert Schweitzer.

The line Carson cited was from a 1956 letter Schweitzer wrote to a friend who expressed concern about pesticides. It was Schweitzer's code red warning for humanity, but it came long before U.N. Secretary-General Guterres issued his. Schweitzer wrote:

"Man has lost the capacity to foresee and to forestall. He will end by destroying the Earth."

Addendum: The Essence of Our Tiny Speck in the Cosmos

I thought the excerpt below from the book, *Pale Blue Dot: A Vision of the Human Future in Space* (Random House 1994) by Astronomer Carl Sagan would be an appropriate addendum for this book which started with the photo *Earthrise* taken by Astronaut Bill Anders.

In February 1990, the space probe, Voyager 1, took a photo of the Earth from a distance of 3.75 billion miles from our planet. Anders' photo was taken from two hundred and forty thousand miles.

Sagan, professor of Astronomy and Space Studies and Director of the Laboratory for Planetary Studies at Cornell University who died in December 1996, offered the following description of the photo (shown on the next page) in his book:

> *Look again at that dot. That's here. That's home. That's us. On it everyone you love, everyone you know, everyone you ever heard of, every human being who ever was, lived out their lives. The aggregate of our joy and suffering, thousands of confident religions, ideologies, and economic doctrines, every hunter and forager, every hero and coward, every creator and destroyer of civilization, every king and peasant, every young couple in love, every mother and father, hopeful child, inventor and explorer, every teacher of morals, every corrupt politician, every "superstar," every "supreme leader,"*

every saint and sinner in the history of our species lived there--on a mote of dust suspended in a sunbeam.

The Earth is a very small stage in a vast cosmic arena. Think of the rivers of blood spilled by all those generals and emperors so that, in glory and triumph, they could become the momentary masters of a fraction of a dot. Think of the endless cruelties visited by the inhabitants of one corner of this pixel on the scarcely distinguishable inhabitants of some other corner, how frequent their misunderstandings, how eager they are to kill one another, how fervent their hatreds.

Our posturings, imagined self-importance, the delusion that we have some privileged position in the Universe, are challenged by this point of pale light. Our planet is a lonely speck in the great enveloping cosmic dark. In our obscurity, in all this vastness, there is no hint that help will come from elsewhere to save us from ourselves.

The Pale Blue Dot, taken Feb. 14, 1990, by NASA's Voyager 1.

The Earth is the only world known so far to harbor life. There is nowhere else, at least in the near future, to which our species could migrate. Visit, yes. Settle, not yet. Like it or not, for the moment the Earth is where we make our stand.

It has been said that astronomy is a humbling and character-building experience. There is perhaps no better demonstration of the folly of human conceits than this distant image of our tiny world. To me, it underscores our responsibility to deal more kindly with one another, and to preserve and cherish the pale blue dot, the only home we've ever known.

Also by Berl Falbaum

Just for Fun, 1972: A book of satirical political one-liners.

The Anchor, Leo & Friends, 1979, JDW Publications, Eugene, Oregon: The true story of a famous bar in downtown Detroit which catered to Damon Runyon-type characters, primarily journalists and politicians.

The Definitive Guide to Organizational Backstabbing, 1994, Avonstoke Press, Troy, Michigan: A satire poking fun at the ethics and culture in the workplace and a sequel, *The Definitive Guide to Organizational Backstabbing, Volume II*, 2007, iUniverse, New York, New York.

A Matter of Precedents, 1998, Proctor Publications, Ann Arbor, Michigan: A novel examining corporate and media ethics.

Shanghai Remembered: Stories of Jews Who Escaped to Shanghai from Nazi Europe, 2005, Momentum Books, Royal Oak, Michigan: In May 2006, the book won an IPPY award from the Independent Book Publishers Association. It was chosen as one of three finalists in the history category.

Prince of Omertá, 2011, Club Lighthouse Publishing, Toronto, Canada: A mobster thriller "ghostwritten" with Giovanni Gambino, of Brooklyn, New York.

Music Is Forever: Dizzy Gillespie, the Jazz Legend and Me, 2014, Red Anchor Productions, Detroit, Michigan: A book about the great jazz trumpeter and creator or bebop, Dizzy Gillespie, and his fifty-year friendship with a white, Jewish man, David (Dave) Usher.

Justice Failed: How "Legal Ethics" Kept Me in Prison for 26 Years, 2017, Counterpoint Press, Berkeley, California: This is the true story of Alton Logan, a black man in Chicago, who served twenty-six years in prison for a murder he did not commit. Four lawyers knew from the outset that Logan was innocent because their client was guilty of the murder; he confessed to them that he had fired the fatal shot that killed a McDonald's security guard. Committed to lawyer-client confidentiality, the attorneys remained silent for twenty-six years until their client died while serving a prison sentence for two other murders of police officers. Falbaum has written a screenplay on the story and, at this writing, was seeking representation from a screenplay agent.

Recollections and Reflections, From a WWII Shanghai Ghetto to Journalism, Politics and Other Journeys along the Way, 2019, S/D Books, LLC, West Bloomfield, Michigan: A memoir of Falbaum's family as refugees in Shanghai after escaping the Holocaust and major stories of Falbaum's career in journalism, politics and PR.

Not One Normal Day, Trumpedia: A Tome of Lies, Scandal, Corruption and Much More (Donald J. Trump's One Term, 2017-2021), 2021, S/D Books, LLC, West Bloomfield, Michigan: A collection of about one hundred and fifty columns on Trump's corrupt administration written by Falbaum for several newspapers in Oakland County, Michigan.

No Comment, a play dealing with media ethics, was produced in 2002 by the Theatre Guild of Livonia-Redford, Michigan.

The Docent, 2013, *The Jewish Magazine* on the Internet: A short story of a Holocaust survivor who volunteers as a docent at a Holocaust museum.

In the late 1970s, Falbaum wrote a manuscript which was highly critical of the media (*The Media and the Public).* In the book, he made numerous recommendations on how to improve the reporting process. It disturbed him immensely that the profession he so loved—the Fourth Estate, so vital to a democracy—was not judged to be trustworthy in countless public opinion polls conducted through the years. Unfortunately, he was never able to have the book published.

About the Author

Berl Falbaum started his sixty-year career in journalism, politics and PR as a general assignment reporter for *The Detroit News* before being promoted to cover local politics, first the Detroit City Council and then the mayor's office as chief of the paper's City-County Bureau. After thirteen years at *The News,* he resigned to become administrative aide to Michigan's lieutenant governor, a position he held for four years.

Then, he served as a speechwriter and media spokesperson for three major corporations before launching his own PR company, Falbaum & Associates, Inc., in 1989. He taught, at night, a variety of journalism courses, including news writing, ethics and law of the press, for forty-five years as an adjunct member of the Wayne State University journalism department. This is his 12th book.

Acknowledgments

William V. (Bill) Haney: I am running out of superlatives to describe Bill Haney. He is a superb author, editor, and publisher, and through the years has given me invaluable advice. Indeed, I probably would not have written any books had it not been for his tutelage when I tried my hand on the first one in 1979. I am extremely grateful and indebted to him.

Anne (Annie) Zimanski: Annie is, without question, the most talented and creative designer I have worked with in my career which covers six decades. As a bonus, there is never a cross word. Whatever the issue, the response from Annie was always, "Sure, no problem; glad to help." For the second time, it was a pleasure working with her.

Carole Cotter Bodner: Carole was assigned the responsibility for giving the manuscript a final proofreading. Given her keen eye, we knew when she said, "It's clean," it would be clean. I am very appreciative for her assistance.

A very special "thank you" to my granddaughter, Jacqueline (Jackie) BeVier. A "grandeditor," she proofed the manuscript and recommended invaluable edits, saving me from some embarrassments. She graduated from the University of Michigan about the same time I finished this project. The best part is that my wife, Phyllis, and I have two other very talented grandchildren, Alex, and Erik.

References

ABC News

Adirondack Council

Alaska's News Service

an inconvenient truth the crisis of global warming

Animal Legal Defense Fund (ALDF)

BirdLife International

Bloomberg

Bloomberg Green

Bloomberg News

Britannica

British Broadcasting Corporation (BBC)

Brookings Institution (The)

Cable News Network (CNN)

Center for Biological Diversity

Chicago Tribune (The)

Citizens for Global Solutions

Climate Council

Climate News

climateclock.world

Climate.gov

climatetippingpoints.info

CNBC

Coastal Care

Conservation International (CI)

Conserve Energy Future

Cornell Lab of Ornithology

Earth Data

Earth.org

EcoWatch

Environmental Defense Fund (EDF)

euronews.green

Forbes Magazine

France 24

Friends of the Earth

Global Climate Change

"Global Temperatures Matter"

Global Vision International (GVI)

Green Matters

Guardian (The)

Half-Earth: Our Planet's Fight For Life

Harvard University

Hebrew Immigrant Aid Society (HIAS)

Hedges, Blair

Impetus Conceptus Pte., Ltd.

Inside Climate News

Institute for Economics and Peace (IEP)

Intergovernmental Panel on Climate Change (IPCC)

Intergovernmental Science-Policy Platform on Biodiversity and Ecosystem Services (IPBES)

International Refugee Assistance Project (IRAP)

International Union for Conservation of Nature (IUCN)

Investopedia

Island Press

Job One for Humanity

KTUU

KYES

Lady Freethinkers

Lifewater

Los Angeles Times (The)

Losing Earth: A Recent History

Mongabay

My Green Lab

National Aeronautics and Space Administration (NASA*)*

National Geographic

National Marine Fisheries Services (NMFS)

National Oceanic and Atmospheric Administration (NOAA)

National Park Service (NPS)

National Public Radio (NPR)

NATO Review (The)

Natural Resources Defense Council (NRDC)

Nature

Nature Geoscience

NBC

New Humanitarian (The)

New Republic (The)

New York Magazine

New York Times (The)

New Yorker (The)

Ocean Agency (The)

Ocean Conservancy

One Earth

Our World in Data

Petpedia.co

Planetary Coral Reef Foundation (PCRF)

Politics Today

Population Media Center

Pop-Up Magazine

Prezi

Pure Michigan

Quartz

Rainforest Trust (The)

Recycle Coach

Reef Revive, Corp.

Rochester Institute of Technology (RIT)

Rolling Stone Magazine

Science

Science Alert

Scientific American

Sea Level Rise.org

Seametrics

Smart Water Magazine

Smithsonian Magazine

Soapboxie

Stimson Center

Surging Seas

Sustainable Air Travel

Temple University

The World Counts

ThoughtCo.

Time Magazine

Trees.org

Uninhabitable Earth: Life after Warming (The)

Union of Concerned Scientists (UCS)

United Nations (U.N.)

United Nations Environment Programme (UNEP)

U.S. Department of Energy

U.S. Energy Information Administration (EIA)

U.S. Environmental Protection Agency (EPA)

U.S. Fish & Wildlife Service (USFWS)

U.S. News & World Report

USA Today

WaterAid

water.org

Weather Channel (The)

"What Climate Change Looks Like from Space"

Wikipedia

Wildlife Conservation Network (WCN)

WION

Wong, Thomas

World Animal Protection (WAP)

World Bank (The)

World Economic Forum (WEF)

World Health Organization

World Vision

World Wildlife Fund (WWF)

worldometer

Yale Climate Connections

Index

Made in the USA
Monee, IL
04 November 2026

67383325R10069